Henry N. Day

Elements of psychology by Henry N. Day

Henry N. Day
Elements of psychology by Henry N. Day
ISBN/EAN: 9783743343337
Manufactured in Europe, USA, Canada, Australia, Japa
Cover: Foto ©berggeist007 / pixelio.de

Manufactured and distributed by brebook publishing software (www.brebook.com)

Henry N. Day

Elements of psychology by Henry N. Day

ELEMENTS

OF

PSYCHOLOGY.

BY

HENRY N. DAY,

Author of "Logic," "Moral Science," "Æsthetics,' "Art of Discourse,' etc.

NEW YORK
G. P. PUTNAM'S SONS
27 AND 29 WEST 23D STREET

PREFACE.

THE governing aim in the preparation of this work has been to furnish a suitable text-book for beginners in metaphysical studies. It has been written predominantly for use in the class-room; and is designed to serve as introductory both to the higher and more critical discussions of the phenomena of mind usually given in the lectures at colleges and universities and also to the study of the derived sciences of logic, ethics, and æsthetics. With these three sciences, which form what has been denominated by Sir William Hamilton Nomological Psychology, its subject matter makes up the entire circle of the mental sciences.

In preparing the work, the general field of psychological literature, as it has been cultivated up to the present time in this country, in Great Britain, and in continental Europe, has been studiously explored, so that all the established results of the most recent investigations might be incorporated into it. This accumulated mass of knowledge, the endeavor has been to reduce into a strictly systematic and scientific form, a form that is indeed the simplest for apprehension by others when attained, but the latest and most difficult of attainment in the progress of science. Something more than accurate presentations of the observed facts of mental action, something more than generalizations of these facts under their appropriate heads, has been aimed at. The endeavor has been to reduce these generalized facts to the exactness of scientific system, in which all the parts are exhibited in their organic interdependence and relation both to the common whole and to one another. The phenomena of mind are thus presented as the manifestations, the affections and the operations, of a single active nature in the diversity of its functions and of its relations to the beings and objects to which it is related. The general attributes of the human mind having been enumerated and explained, the particular phenomena of mental activity, the facts of sensibility, intelligence, and will, are exhibited as the states of a single active nature which, while revealing more prominently and characteristically sometimes this and sometimes that side of its composite life, never wholly drops out of its phenomenal action any constituent element of its being.

The particular phenomena of mind as classified subjectively by psychologists generally at the present time under the departments of intelligence, sensibility, and will, are in this work treated in the light of their respective relation and correspondence to the old and still unquestioned classes of phenomena handed down to us from antiquity under the objective enumeration of the True, the Beautiful, and the Good. Psychological science, in the light of this correspondence, it is believed, is enabled to exhibit its phenomena in a new clearness and impressiveness.

The order of treatment, hitherto adopted, has been in this work varied by giving priority to the department of the sensibility before that of the intelligence. This is unquestionably the order of natural manifestation, as we must have sense of an object—must be impressed by it—before we can think of it. This natural order, as might be anticipated, prevents much obscurity and confusion and consequent error in the explanation of certain mental states, particularly those of the imagination and memory. These two states have been generally, and of course very erroneously, presented under the intelligence or cognitive function.

Further, the department of the sensibility has been treated with more fullness and more scientific method than has been usual heretofore. This department has been far less cultivated than the departments of the intelligence and the will. Yet even these functions are so closely and vitally related to that of the sensibility that they cannot be fully and accurately shown except in their relation of interdependence to the sensibility.

This elementary work in mental science is contributed to our text-book literature, in the earnest hope that it may be found to be serviceable in some degree to the elevation of the study of mind to its true and proper rank in the circle of educational studies.

NEW HAVEN, February, 1876.

CONTENTS.

INTRODUCTION.—§ 1. DEFINITION OF PSYCHOLOGY.—§ 2. SOURCE OF KNOWLEDGE.—§ 3. PROVINCE OF PSYCHOLOGY.

BOOK I.

GENERAL ATTRIBUTES OF MIND.

CHAPTER I.—THE ESSENTIAL ACTIVITY OF MIND.—§ 4. The Mind essentially Active. § 5. Definition of Mind. § 6. Three forms of Mental Activity.

CHAPTER II.—THE SINGLENESS AND SIMPLICITY OF MIND.—§ 7. The Mind Single and Simple. § 8. Each Mind a Distinct Unit. § 9. Mind not the same as its Object. § 10. Mind Homogeneous.

CHAPTER III.—THE FINITENESS AND DEPENDENCE OF THE HUMAN MIND.—§ 11. The Human Mind Limited and Dependent; —in its Range. § 12. In its Intensity. § 13. In its Dependence on Objects. §§ 14-16. The Dependence of Mind on its Objects, Threefold.

CHAPTER IV.—THE PASSIVITY OF THE HUMAN MIND.—§ 17. The Human Mind Passive. § 18. The Mind Passive in Respect to Objects Without. § 19. In Respect to its Own States. § 20. Impressions on the Mind Vary. § 21. Mind Active and Passive in Every Experience. § 22. Pleasures from its Activity and Susceptibility.

CHAPTER V.—THE CONTINUOUSNESS OF MIND.—§ 23. The Mind Continuous in its Activity and its Passivity. § 24. Habit. § 25. Mental Growth.

CHAPTER VI.—THE SELF-CONSCIOUSNESS OF MIND.—§ 26. The Mind Conscious of its own Acts and Beliefs. § 27. Self-Consciousness Respects not the Mind Itself but its Modifications. § 28. In Self-Consciousness, the Mind both Knows and Feels. § 29. Self-Consciousness Varies in Degree.

CHAPTER VII.—THE RELATIVENESS OF MIND TO ITS OBJECTS.—IDEAS.—§ 30. The Mind Knows Only in Relation to its Objects. § 31. Ideas, What. § 32. Nothing but Idea Object to the Mind. § 33. Nothing but Mind Subject for Idea. § 34. Twofold Relation between Mind and Idea. § 35. Idea More Exactly Defined § 36. Any Expressed Idea Apprehensible either as True, or as Beautiful, or as Good. § 37. The True. § 38. The Beautiful, § 39. The Good. § 40. The True, the Beautiful, the Good, Respectively Objects for the Intelligence, the Sensibility, and the Will.

CHAPTER VIII.—§ 41. Symbols of Mind.

BOOK II.

THE SENSIBILITY.

CHAPTER I.—ITS NATURE AND ITS MODIFICATIONS.—§§ 42-44. The Sensibility Defined and Explained. §§ 45,46. Form. § 47. The Imagination. § 48. The Two Departments of the Phenomena of the Sensibility. §§ 49-54. The Feelings Distinguished into Classes.

CHAPTER II.—PLEASURE AND PAIN.—§ 55. Pleasure and Pain Defined. § 56. Distinguishable in Thought from both act and feeling. § 57. Degrees. § 58. Attendant on Every Mental Experience. §§ 59-63. Modifications.

CHAPTER III.—THE SENSATIONS.—§ 63. Sensation Defined. § 64. Referable Solely to Mind. § 65. Medium of Sensation. § 66. The Nerves,—Two Classes. § 67. Sensation the Same, whatever Part of the Nerve is Affected. § 68. Sensations Classified § 69. Simple Bodily Pleasure and Pain. § 70. General Vital Sense. § 71. General Organic Sense. § 72. Special Sense. § 73. Touch. § 74. Taste. § 75. Smell. § 76. Hearing. § 77. Sight.

CONTENTS.

CHAPTER IV.—THE EMOTIONS.—§ 78. Emotion Defined. §§ 79,80. Three General Classes of Emotions. § 81. Intellectual Sense. § 82. Aesthetic Sense,—Beautiful, Sublime, Comic. § 83. Moral Sense.

CHAPTER V.—THE AFFECTIONS.—§ 84. Affections Defined. § 85. Love and Hate. § 86. Habitual Dispositions. § 87. Affections in Relation to Their Objects. § 88. Resentments.

CHAPTER VI.— THE DESIRES.— §§ 89,90. Desires Defined and Characterized. § 91. Aversions. § 92. Desires Classified in Reference to Capability of the Soul. § 93. Appetites. § 94. Rational Desires. § 95. Social Desires. § 96. Hopes and Fears.

CHAPTER VII.—THE SENTIMENTS.—§ 97. Sentiments Defined. § 98. Sentiments Classified. § 99. Contemplative Sentiments. § 100,101. Practical or Moral Sentiments.

CHAPTER VIII.—THE PASSIONS.—§ 102. The Passions Defined and Classified.

CHAPTER IX.—THE IMAGINATION.—§ 103. The Imagination Defined. § 104. Diverse Denominations of the Faculty. § 105. The Nature Illustrated. §§ 106-108. Ideals Defined and Classified

CHAPTER X.—THE IMAGINATION.—SENSE-IDEALS.—§ 109. Sense-Ideals Defined. §§ 110,111. Sense-Ideals, as Related to the Sensuous Organism and to the Mind. § 112. Phantoms. § 113. Exalted Sensibility. § 114. Suspended Sensibility. § 115. Dreaming. § 116. Catalepsy. § 117. Somnambulism.

CHAPTER XI.—THE IMAGINATION—SPIRITUAL IDEALS;— § 118. Spiritual Ideals Defined. § 119. Spiritual Ideals Analyzed and Exemplified.

CHAPTER XII.—MEMORY.—§ 120. Memory Defined. § 121. Its Law. § 122-126. Conditions and Special Rules of a good Memory.

CHAPTER XIII.—MENTAL REPRODUCTION.— § 127. Mental Reproduction Defined. § 128. Mental Reproduction, Spontaneous or Voluntary.—Revery. §§ 129-135. Laws of Mental Association. § 136 Voluntary Reproduction. §§ 137-140. Rules of Recollection

CHAPTER XIV.—THE THREEFOLD FUNCTION OF THE IMAGINATION.—§ 141. The Three Functions of the Imagination

Enumerated. § 142. The Artistic Imagination—Æsthetics. § 143 The Philosophical Imagination.—Logic. § 144. The Practical Imagination.—Ethics.

BOOK III.
THE INTELLIGENCE.

CHAPTER I.—NATURE AND MODIFICATIONS.—§ 147 Intelligence Defined. §§ 148–151. Its Modifications.

CHAPTER II.—PERCEPTION.—§ 152. Perception Defined § 153. Relations to Sensation. § 154. Sphere of Perception. §§ 155–157. Kind of Knowledge given by Perception.

CHAPTER III.—INTUITION.—§§ 158–159. Intuition Defined. § 160. Sphere of Intuition. §§ 161, 162. Kind of Knowledge given by Intuition.

CHAPTER IV.—THOUGHT.—§§ 163, 164. Thought Defined. § 165. The Three Essential Elements. § 166. Thought or Act of the Discursive Intelligence. § 167. Its Three Forms. § 168. The Judgment. § 169. The Concept. § 170. The Reasoning. § 171. Classes of Attributes. § 172. Intrinsic Attributes—Qualities and Actions. § 173. Essential and Accidental Properties. § 174. Attributes of Relation and of Condition.

CHAPTER V.—THE CATEGORIES OF THOUGHT.—§§ 175, 176. Category Defined. § 177. The Category of Identity and Difference. § 178. The Category of Quantity. § 179. The Category of Modality. § 180. The Category of Properties and Relations. § 181. Categories of the Beautiful, the True, and the Good.

CHAPTER VI.—EXISTENCE.—§ 182. The Reality of Objects. § 183. Mind. § 184. Matter. § 185. Universe. § 186. Substances and Causes. § 187. Space and Time.

CHAPTER VII.—INTELLECTUAL APPREHENSION AND REPRESENTATION.—§ 188. Intelligence as Capacity and as Faculty § 189. Intellectual Apprehension. § 190. Intellectual Representation.

CHAPTER VIII.—CURIOSITY AND ATTENTION.—§ 191. Intelligence Instinctive or Voluntary. § 192. Curiosity. § 193. Attention.

BOOK IV.

THE WILL.

CHAPTER I.—NATURE AND MODIFICATIONS OF THE WILL.—§ 194. Will Defined.

CHAPTER II.—CHOICE.—§§ 195-199. Choice Exemplified and Explained. §§ 200-203. Free Personality Analyzed.

CHAPTER III.—MOTIVE.—§ 204. Motive Defined. §§ 205, 206. Motive ever a Good to the Whole Soul and in the Mind. §§ 208, 209. Motives Classified as External and Internal.

CHAPTER IV.—GROWTH AND SUBORDINATIONS OF WILL.—§§ 210, 211. The Will a Power Capable of Growth. § 212. Dependence of Will. § 213. Governing and Subordinate Volitions.

CHAPTER V.—CONSCIENCE.—§§ 214, 215. Conscience—Its Motive and Elements. § 216. Discernment of Right and Wrong. § 217. Sentiment of Obligation. § 218. Sense of Approval. § 219. Conscience Influenced by Will.

CHAPTER VI.—HOPE, FAITH, AND LOVE.—§ 220. Hope, Faith, Love, as Virtues. § 221. Hope Defined. § 222. Faith Defined. § 223. Love Defined.

INTRODUCTION.

Psychology defined.
§ 1. PSYCHOLOGY is the science of the human mind.

The term *psychology* is from the Greek language, and properly signifies a discourse upon the soul or mind.

As a science, psychology professes to set forth the facts pertaining to the human mind or soul in logical order and completeness.

The terms *mind*, *soul*, and *spirit* are all used to denote that in man which thinks, feels, and wills. They are used, however, often with different shades of meaning. The word *mind* points rather to the thinking faculty, the intelligence ; *soul* indicates more immediately the feeling capacity, the sense ; while *spirit* rather regards the rational nature that wills and directs its own acts. Other synonymous terms are *conscious subject, the self, ego, the human consciousness*.

Sources of knowledge.
§ 2. The one source from which we obtain our knowledge of the facts of mind is experience. This experience is either our own or that of others as shown in their expressed acts.

As we shall see, the more essential facts of mind

are its operations, because it is essentially of an active nature. We are able to take notice of the operations of our own minds. Men of reflection in all ages have sought to know themselves, their feelings, their thoughts, their endeavors; the various kinds of their mental operations; and the relations of those acts or states to one another and to their respective objects. These observations and studies have been recorded; have been compared; and thus the world has ever been increasing its knowledge of the mind. Moreover, men have framed language, have introduced words and shaped sentences, so as to set forth more exactly their thoughts. Thus it has come to pass that the language of men is a rich treasure-house of information in regard to the facts of mind. The literature of the world which contains the record of what man has done, what he has experienced, what he has observed in himself and the world around him, what he has thought and suffered and achieved, is the grand repository of knowledge in regard to the nature and the attributes of the human mind.

In fact, it is in language, in the varying significations of words, which must vary to mark the progress of knowledge, that we find not only the best light in the study of mind, but also the occasion of the most serious mistake and error and dispute, just as a changing, intermitting light may mislead, even although it be the only guide we can obtain.

Province of psychology. § 3. From these facts—these *phenomena*—of mind, gathered from the universal experience of men through personal observation and study of the experience of others recorded in language and literature, we are enabled to discern the general characters or attributes of mind and thus far to know its nature. It is accordingly the particular province of psychology to set forth in order and completeness the general characteristics or attributes of mind.

All true knowledge is in fact but knowledge of attributes. All our knowledge of *the sun* consists in our knowing certain attributes belonging to it, as that it is round, is bright, is in the heavens, is centre of the planetary system, and the like. Of the sun as a substance, apart from its attributes, we know nothing; at least we know nothing more than this, that it is not the moon, is not the earth, is not any thing else that we can name. All positive knowledge terminates in attributes. When we know all the attributes of an object, we know everything that can be known in relation to it; while we gain no knowledge of it whatever, until we recognize some attribute belonging to it. All science, accordingly, searches for attributes; deals with attributes; and deals with nothing else. Psychology, as a science, consequently, seeks only to recognize the attributes of mind, enumerate them all in their proper order, and exhibit them in their respective relations to one another, and to their several objects.

We have thus the proper method of studying the mind prescribed to us. We are to gather up the facts of mind, for the purpose of finding all the general attributes of mind. We are then to exhibit these attributes one by one in such way as to show that our survey of these attributes is complete, and that no one is omitted, since this assurance of completeness it is the very object of a science to give. We are to exhibit them, moreover, in their true relations to one another; and, also, inasmuch as the mind will be found to be but part of the vast whole of creation, in their true relations to other things so far as they may become objects to it.

Method. Our method, accordingly, will lead us, in the first part of our study, to exhibit the general attributes of mind, and in the three subsequent parts respectively, to exhibit in order the special attributes of the mind as essentially of an active nature, as affecting other minds and objects, and as equally affected by them.

BOOK I.

GENERAL ATTRIBUTES OF MIND

CHAPTER I.

THE ESSENTIAL ACTIVITY OF MIND.

§ 4. The mind is essentially active.

<small>Mind essentially active.</small> As we survey the facts of mind—its phenomena—we discover everywhere an active nature. There is not a fact which does not reveal this activity. Our thoughts are active; our purposes are active; our feelings even are the modifications of an active nature. The feelings of the human soul are not like impressions on an inactive substance as on a stone. So essential is activity in them that they are in a certain truth sometimes reckoned as the active powers of the mind.

It is by this property of activity that mind is distinguished from matter, which is the only other kind of being of which we have any knowledge. Matter is inactive, inert; it moves only as it is moved, and just as it is moved; it is moved only by mind as the only moving force of which we know anything. Matter is not only motionless but also formless, except as it is moved and shaped by mind.

The best definitions that can be given of mind

are accordingly those which define it through this attribute.

Mind defined. § 5. If we define mind by distinguishing it from all other being, we have the definition: MIND IS THAT KIND OF BEING WHICH ACTS; while matter, the only other kind of being, is that which is inert or incapable in itself of action. Or, if we define it by its characteristic properties, we have the definition: MIND IS THAT WHICH THINKS, AND FEELS, AND WILLS.

This then is the fundamental and essential attribute of mind—activity. As we shall see, all other attributes are really but modifications or characteristics of mind as active. So that wherever there is mind there is activity, as the essential element.

Activity peculiar to mind. Not only is activity the essential attribute of mind, but it is peculiar to mind. There is no activity that we know of which is not the activity of mind, or the proper and sole result of its activity. On the other hand, wherever there is activity, there we discover mind immediately or remotely revealed. All causes are but modes of mental activity. And as we, instinctively as it were, demand for all things a cause, and if we recognize second causes, still demand a first great cause, and can conceive of such cause only as a determination of some mind; so we are forced to regard mind as the only activity of which we can conceive. At least, psychology, which is a science of observation, recognizing the fact that mind does originate action,—that its purposes act upon thoughts and feelings, on other minds, on matter even, moving the limbs and other organs of the body and through them even material objects without,—recognizing this fact and observing no motion or effect which it cannot account for on the ground that it is the immediate or remote effect of the action of mind, does not regard itself as called upon to suppose any other activity. Until proof is brought to it, this science cannot in its teachings recognize that there is any activity other than that of mind.

§ 6. There are to be recognized three very different forms or kinds of mental activity; in other words, we observe three different functions of the human mind. If an orange be put into the hand, the mind at once feels certain impressions made upon it ; there is a feeling of softness, of roundness, of agreeable perfume. The mind feels from some object out of itself. One function of the mind then is feeling. This, its function of feeling, is called the *sensibility*. The mind also through these senses of touch and smell and still more through that of sight, perceives this exterior object ; it knows it to be there in the hand, it knows it to have these attributes of form and smell and color ; the mind thus perceives and knows ; and this, its function of knowing, is called the *intelligence*. Moreover the mind determines to hold fast the orange, to move it before its eye, to smell or taste it ; this is another form of its activity ; and this, its function of determining, is called the *will*.

<small>Three forms of mental activity.</small>

We can discover no other form of mental activity ; we can conceive of no other. These three forms are equally necessary each to the others. We cannot know or choose the orange, until the orange has been introduced to our intelligence and our will through some impression upon us, through some feeling: and all feeling in man is more than mere impression on a blindly passive, immovable rock; it is impression on that which knows it feels and which determines and acts; as for instance, by throwing off the impression, or by allowing it and then permitting its thought to rest upon it and then perhaps determining upon some act in relation to it.

These three great functions of the mind presenting three different departments of mental activity or three distinct classes of mental phenomena, mark out the three grand divisions of psychology which treat respectively of the three specific forms of mental activity or the three special functions of the human mind.

CHAPTER II.

THE SINGLENESS AND SIMPLICITY OF MIND.

Mind single and simple.
§ 7. The human mind is single and simple.
It is single in the sense that it is one by itself, separate from all other beings. It is simple in the sense that it is homogeneous in its whole nature and cannot be decomposed into separate elements.

Unity of mind.
§ 8. Each mind is a unit, by itself, existing separately from all other minds and from all other beings. This implies that there is a plurality of minds and that each individual mind is one of a number. Psychology, as a science, assumes this as a fundamental fact.

It belongs to another science to discuss the general question in all its bearings, whether there is more than a single mind in the universe of being. That there is but one mind, and that consequently each human mind is an inseparable part of this one universal mind, is one form of pantheism—that of idealism—which denies the existence of matter as distinct from mind; the other form of pantheism being that of materialism, which denies the existence of mind as distinct from matter.

Proved from experience. Psychology, which does not properly inquire into the true significance and proof of these pantheistic doctrines, satisfies itself with the test of observation and experience. Every one knows from himself, that he is not a mere mode of existence of some one universal being. He knows of himself that the life of his mind is as distinct from that of other minds as the life of his body is distinct from that of other bodies; that he thinks, and feels, and wills, not as a part of a universal thinking and feeling and willing, just as much as he breathes and has a sense of pain or pleasure and moves his limbs by himself and not as a part of a universal animal life. He feels a responsibility for many of his actions, because he knows they are his own and not those of a being called in the language of theory and speculation the absolute and the infinite. All the records of human experience unite their testimony in confirmation of what each individual knows in his own consciousness. Only in figurative language, used to help out the notion of the close relation between the human mind and the divine mind, can it be said, that there is one sole mind in the universe, one function of thinking, one of feeling, one of willing. There is a multitude of minds; each human mind is one of this multitude, existing by itself.

Mind distinct from its object. § 9. The human mind is not the same as the object with which it has to do. This is another form of speculation which has arisen in the endeavor to account for the communication between the mind and the object of which it thinks, which it feels, which it chooses. That thought and its object are one, is the unwarranted assumption of some theorists. It is enough here to reply, that this supposition is contradicted by the testimony of each man's consciousness as well as by the records of the experiences of men generally. When I think of the sun, I know that the sun is different, is not one with myself, in any but a figurative sense. I may, in a warrantable use of language, say, when I think of the sun, that the sun is in my thought, is in my mind; but this language is strained beyond its true meaning when it is construed to imply that the sun is not external to me, and truly without my mind and as an existence separate from it.

Simplicity of mental action. § 10. The human mind is homogeneous and cannot be decomposed into simpler elements.

It has in a certain sense, divers functions; but it is the same mind that acts in each. If we distinguish the several functions of thinking, of feeling, and of willing, it is still the same one mind that is engaged in each. It is the same one river that yields to the ship that is launched into it and that bears it down its current; that hardens its surface under the frost and scatters spray when it dashes upon a rock. It is the same wind that chills us and drives us. We cannot separate a river into something that shall simply yield to a ship and something else that shall float it down its current; or a wind into something that shall chill us and something else which shall beat upon us and drive us, which two things may be combined and make a wind. So we cannot separate the mind into three several things, one of which can think, another feel, and the third will. On the contrary, it is inconceivable how the mind can think without feeling or without willing.

CHAPTER III.

THE FINITENESS AND DEPENDENCE OF THE HUMAN MIND.

Human mind limited in range.
§ 11. The human mind is limited and dependent.

It is limited in its range. It can reach but a part of the universe of objects.

The range of its activity may enlarge with the expansion and growth of which its nature is capable through interminable ages, and yet a boundless universe will ever infinitely outstretch its largest capacity. More than this, the range of its activity varies during this progress and growth. There are times in which the mind is conscious that it has not the power which it has at other times. It feels itself at such times shut up within narrow limits, as if by a power without, so that it cannot exert itself with the grasp and energy that it knows truly belongs to it.

In intensity.
§ 12. The human mind is limited also in its intensity.

The force with which it acts may increase indefinitely, but it never can become comparable to the force of the infinite mind. In infancy and childhood it is very imbecility and weakness; in adult life, it is relatively strong and vigorous; yet while we anticipate from the very instincts of our nature an ever advancing increase of power, this instinctive anticipation never looks forward even to a near approach to the power of the infinite mind. So, too, it is variable in respect to the strength already in a sense acquired. At one time it is listless, dull, heavy, even as compared with what it had been a few hours before. At another time it is comparatively quick, bright, and vigorous; it observes, and judges, and feels, and acts with far greater promptness and keenness and force.

§ 13. The human mind, moreover, is limited in being dependent on its object for the exercise of whatever range and intensity of capacity actually belong to it. It cannot act at all without an object.

Dependent on object.

After some experience, indeed, its own thoughts and feelings and endeavors may be objects to it. Upon these it may exercise its various functions without looking out of itself. But at first absolutely, and ever to a great extent, it depends on objects out of itself. For the most part these objects are out of its reach. They must be brought to it.

§ 14. The dependence of the human mind on its objects is of a threefold character.

First, the mind is dependent on its needed objects coming to it.

1. Presence of object.

At the first, as the mind is without any experience, it cannot know what objects there are for it to act upon, or where such objects are to be found. It knows not even that it is to act or can act at all, for it knows literally nothing. The first activity is to be awakened and called forth by some object being brought to it. That first object presented to it comes to it by no agency of the mind at all, for it knows nothing of there being any such object till the object comes to it and awakens it to active life. Consequently, what particular one of its several functions, what department of its active nature, so to speak, is to be addressed, it cannot determine for itself, for it knows nothing of object or function. And so ever afterwards, objects come to it without its choice, and engage its activity in forms which not itself, but the objects themselves, in a great measure, if not wholly, determine.

The universe of objects around it, is disposed and moved by a power without itself, and altogether beyond its control.

§ 15. Secondly, the human mind is dependent for its objects to a great degree upon channels or means that are not under its own control.

2. Channels and means.

It is thus dependent on the thousand channels and means by which objects are introduced to it. But we need here only instance that wonderful assemblage of channels and organs compacted together in the human body. These organs which we term the senses, one or the other of them, convey to the mind its first object and afterwards all the new objects about which it acts. The light of the sun comes in vain to it, except for the organ of the eye through which alone the mind is reached. The music which fills the air around gives no sense of melody except for the ear which conveys it from without. Over these senses, the mind has, before they have awakened its first experience, of course no control, since, by the supposition, it has till then no knowledge; and afterwards, while it may use and direct them to a limited extent, yet its power over them is extremely limited. It cannot hinder disease from weakening or destroying them; it cannot prevent their becoming weary and dull. They are, moreover, in themselves limited organs. They can reach out few objects comparatively in the vast universe; they can of these accessible objects convey but a few at a time.

§ 16. Thirdly, the human mind is limit-
3. Control of object. ed in its control over its objects when actually brought to it.

These objects often force themselves upon it and domineer over its thoughts, its feelings, and its endeavors. Perhaps the mind never more sensibly feels its weakness and its dependence than when it strives to banish from its thoughts a disagreeable object. A wrong action, an ill-timed word, a mistaken step, which causes shame and remorse, how it will fasten itself upon the mind, disturb its rest, haunt its dreams, poison every pleasure, in spite of the mind's utmost effort to expel it.

Thus dependent is the human mind. Its active nature must have objects to go out upon. But they are out of its reach and are brought to it through channels and organs beyond its control, and then often when they are brought to it against its will, it lies helpless, without strength or skill to banish them or to hinder their return.

CHAPTER IV.

THE PASSIVITY OF THE HUMAN MIND.

§ 17. *The human mind is passive.*

<small>Passivity of mind.</small> The mind receives impressions from the multitude of objects around it. It comes in contact with them at every turn. It cannot avoid them altogether. Indeed, it must have been impressed by them in some way before it could know that they are to be avoided. Only as it is able to put itself out of the universe of God can it escape impressions from the objects around it.

Moreover, as we have seen, the mind cannot act itself unless thus impressed. It must be passive in order to be active. The first act is called forth by some object impressing it. And ever after, the indispensable condition of any exertion of its activity is its receiving some impression from some object.

§ 18. *The mind is passive or suffers* <small>Twofold.</small> *impressions both from objects out of itself and also from its own states.*

The first impression is ever from some object external to itself. And through the whole <small>1. From without.</small> course of its history, new objects are perpetually coming in, some sought, some unsought, some acceptable, some disagreeable, and making each its own impression upon it.

§ 19. The mind, too, suffers impressions from its own states. A memory of past dangers impresses it, and from the impression a feeling of gratitude, a thought of wise precaution, a purpose of self-protection, arise in it. It feels an emotion of anger; and the feeling reacts and impresses it with a sense of shame. It thinks out a new discovery, and the thought impresses it with exultation. It resolves on some noble deed, and the purpose reacts and impresses it with joyful self-complacency. Thus one thought, impressing the mind as passive to such impressions, becomes the occasion of a second thought; and this of a third; and so on. One feeling in like manner reacts on the passive mind, and impresses a second feeling, or gives strength to the first, and this feeling becomes ever the occasion of new feeling. It is the same way with purposes. They react and impress the passive mind according to their own character, or according to the condition of the mind itself at the time.

2. From within.

§ 20. The impressions on the mind vary with the varying states of the mind. At one time the mind is more passive, more susceptible of impressions. In infancy, it has little control over itself, and takes just what impressions the object is fitted to make. The adult mind has acquired the power both to resist impressions to a certain degree, and also to determine within certain limits what kind of impression shall be made upon it. An angry blow, which would provoke unavoidable resentment in a child, may, in the trained and cultivated soul, only call forth

Varying with its own states.

forgiveness and beneficence. But in any one of the great stages of life on different days, or even at different hours, the soul is variously affected by the same object, on account of the varying condition of its own experience.

§ 21. The mind, accordingly, while essentially active in its nature and thus ever exerting its activity in relation to some object, is also passive in so far as receiving and retaining that object before it. It is thus both active and passive at the same time.

<small>Active and passive at the same time.</small>

"There is no operation of mind," says Sir William Hamilton, "which is purely active; no affection which is purely passive. In every mental modification, action and passion are the two necessary elements or factors of which it is composed. But though both are always present, each is not, however, always present in equal quantity. Sometimes the one preponderates, sometimes the other."

When action preponderates so as to characterize the mental state, we speak of the mind as a *faculty*; when passion or feeling predominates, we speak of it as a *capacity*. We have thus a faculty of knowing truth and a capacity of receiving truth or knowledge; we have a faculty of creating beautiful forms, and we have a capacity of feeling beautiful forms; we have a faculty of doing good; we have a capacity of feeling kindness. A mental faculty is the mind's ability to act in some way; a mental capacity is the mind's ability to feel in some way.

§ 22. Not only is the human mind ever receiving impressions from objects without, and also from its own acts and states

<small>Activity always attended with pleasure.</small>

becoming so far objects to it, but it also experiences a pleasure from its very activity and susceptibility when in legitimate exercise.

There is thus a pleasure in knowing, even aside from the character or nature of what is known. There is a kind of satisfaction in hearing bad news; there is pleasure in knowing that which has not felt good in itself to us. We take a kind of satisfaction too in witnessing tragic scenes. Suffering and distress interest us, even while they give us pain. Aside from the objects themselves as good or evil to us, there is a natural pleasure in our actions and our impressions in relation to them. We discover here a ground for the truth of the familiar saying that sympathy doubles joys and divides griefs. We see here also why we love life and shrink from the thought of its coming to an end.

This fundamental pleasure attending the legitimate acts and affections of our minds is enhanced when these acts and affections are felt to be in harmony with their objects. If the objects of our study are exactly accordant with the activity and maturity of our intelligence; if they are recognized to be in harmony with the truths already in our minds, a higher, broader satisfaction is felt than that which attends the mere exercise of our minds.

Not only must we presume from the perfect goodness of the Creator that the legitimate use of all the powers with which he has endowed us must be good to us, must give us pleasure or joy, but in addition to this, we must presume from his infinite perfection that the universe around us must be in harmony with our natures, and, therefore, that the experience of this harmony between ourselves and the world of being and of events about us should be a source of joy to us. In this pure satisfaction and pleasure which fills our souls when we think and feel in respect to these objects, we find a test of truth, of beauty, and of moral goodness or rectitude. That can hardly be accepted to be true which when known gives us in the knowledge no satisfaction. The pleasure felt in beautiful objects is so prominent a part of our experience that some philosophers have made this the sole test of beauty in an object, that it gives us pleasure. In like manner the satisfaction of the moral sense is accepted as one leading test of what is right and morally good.

CHAPTER V.

THE CONTINUOUSNESS OF MIND.

Menta. activity continuous.

§ 23. The human mind is continuous in its activity and in its passivity.

As it is essentially active, were it to cease its activity, it would lose its essential attribute and consequently cease to be. In the same way, were it to intermit its capacity to receive or retain an object for its activity, it must cease to be.

We cannot, accordingly, conceive of a mind unless as during its existence continuing to be affected by some object either external or internal arising from past experience and also continuing to act in relation to some object.

Further, we have every reason to presume from its present acting and feeling, that it continues to act and feel during its entire existence ; for, as an existing power to act and feel, it has ever continuing to it all the necessary conditions of acting and feeling in the universe of objects about it. Only by passing out of this universe can it be placed beyond these conditions. When the power to act exists and all the conditions of its acting continue to be supplied to it there is a necessity of its acting.

Moreover, there is no good reason for supposing that it ever during its existence wholly intermits its acting or its feeling. It is in sleep perhaps, that the mind, if ever, ceases to act or intermits its actions. It is true indeed that we cannot, when we awake, recall but little of any thinking or feeling that we have experienced while we have been asleep. But surely our inability to recollect what has passed in our minds at any time is not decisive proof that we did not think at all at that time. Many hours are passed when we are awake and our minds are known to be active, and yet after a day, or it may be after a few hours, we are unable to recall a solitary thought that was in our minds during the time. It is often the case, indeed, that when we are asked what we have been thinking of during the previous hour we find it difficult to tell much, and impossible to recollect all.

Many facts, still further, prove the continuous, unintermitting activity of the mind during sleep. When we wake, we very often notice that our minds are occupied with a train of thoughts which had commenced during sleep. The often observed restlessness of persons while asleep show that their minds are agitated by some subject. We naturally and legitimately infer that it is only because the thoughts are not of so exciting a nature that they do not disclose themselves in more tranquil sleep.

The remarkable facts of sleep-walking—Somnambulism—show us that the mind may act through a period of several hours in incessant and very energetic activity, and yet be unable, when it is over, to recollect anything that has occurred during the sleep. These facts, which are well accredited, are such as working out difficult problems, executing nice processes of art, directing the feet securely along very difficult passages, performing feats of mind and of body beyond all that is possible in wakefulness. Somnambulists are nevertheless generally unable when the sleep is over to recall any thing that has happened during those hours of disturbed slumber.

<small>Ground of personal identity.</small> This never intermitted continuousness of mental activity enables us to know that we are the same beings to-day that we were yesterday. It furnishes the sufficient and only decisive evidence of our personal identity. If a link were to fall out in this continuous chain of our mental life, the chain would break into parts that we should be wholly unable to reunite. There would be a chasm that the mind could not recross. All its thoughts and feelings before the interruption must remain to it strange and foreign; the acts of another being than itself. It is because and only because I can send the electric thought of my present soul back along the successive, undissevered links of my past thought to the thought of yesterday, and of last year, of any past epoch of my life, that I know that thought to be the thought of my present self and myself to be the same I who think now, that thought then and thought that thought.

This image of an unbroken chain is inadequate to the full truth of the case. The thought of yesterday is itself in my mind in a true sense to-day. I am different to-day from what I should be but for that thought; my mental life is changed by it. If I could be conscious of the minutest affections of my soul, if with omniscient ken, I could discern the entire body of my present thought, I should discover that thought of yesterday still living within the body of my present living thought. We know that thoughts which greatly interest us, strong feelings, important purposes, do thus reach along and live along in us for days or even years. There is nothing certainly in the nature of the more trivial thoughts or feelings or purposes of our minds, so different from these more notable experiences, which forbids the supposition that they also live on as well; only our finite minds are unable to discover them. It is thus I know myself to be the same being I was yesterday, a year ago, by knowing that the thoughts of yesterday and of last year are still in my mind, the same thoughts, however modified by subsequent experience, however differently related to other thoughts in my mind, the same thoughts still. Because they are the same, I who thought and think then am the same.

§ 24. This continuousness of mental life is the ground and essential element of *habit*.

Habit defined.

HABIT is the holding on of the mind in any kind of acting or feeling. Habit presupposes some first starting or origination of an act or feeling from some cause, and denotes the continuance of the act or feeling without necessary repetition of the act of causing or originating. It belongs to every department of our mental life. A particular kind of feeling, the more it is allowed, the more prevails, the more easily returns. The mind becomes more yielding to the object of it, more impressible by it. The organs of sense—the nerves of the eye or of the ear—may become more and more dull, but the mental sensibility becomes more ready and yielding. The thought that has once been awakened in the mind comes more easily on the second occasion. It prevails more and more as it is more frequently repeated, till at length it bears itself along as it were by its own acquired strength; it holds on and usurps a place in the experience, even crowding out perhaps other thoughts that seek to enter and engaging the mental activity more and more with itself.

Memory.
This habit of mind in regard to thought, in regard to everything which we know or of which we are conscious, is the main element in memory. This is indeed all there is in simple memory as distinguished from recollection; a simple holding on of a thought in the mental life, or the holding on of the mind to a thought it has once experienced.

In feelings.
Habit, as already observed, attaches as well to feeling as to thought. A particular feeling allowed repeatedly becomes a passion ruling sometimes over all the faculties of the soul. It springs up into prominence on the slightest occasion. A word, even a memory, brings it to the surface of the consciousness. It lives on, defying our best effort to extirpate or crush it. Feelings of cheerfulness, of anger, of pity, of confidence, and the like, come to be so habitual, to appear so readily and surely on every occasion, that the

character is marked by them and the man is recognized by all as one in whom these dispositions reign continuously.

In purposes. Habit attaches likewise to purposes, to all voluntary acts. What we do often, we come to do almost or quite unconsciously. We purpose to take a walk; the purpose holds on and bears us along over long distances, without ever coming up again into distinct consciousness. We take step after step, we shun obstacles, we turn corners; the purpose holds on through all, and we take no heed of it unless arrested in our way by some obstacle or some diversion, as by some friend meeting us, or by some new object or occupation coming into our minds. In the same way dexterity and skill are acquired by this holding on of purpose, directing the movements of our hands, our fingers, or our feet. In order to touch the keys aright for playing a given melody there is at first a necessity for a special purpose or volition to move the fingers for each separate note. As the tune is played over and over, the fingers come to move as if of themselves. We are conscious of no special purpose or volition; but it must yet be there, governing every movement. It holds on and as new energy is imparted to it by every repetition it becomes finally strong and quick and executes rapidly, accurately, skillfully. This continuousness of purpose, this habit of endeavor, is the condition of all right and good character. Only as the past purpose lives on, can there be anything, in the intention at least, to make the future like the past; and life and action would without this continuousness be a succession of unrelated purposes and acts determined not by an inner principle but by outer occasion, so that there could be no such thing as proper character.

Condition of growth. § 25. This continuousness of mental activity and passiveness, once more, is the condition and chief element in all mental growth.

Everything that lives and acts may grow in two different ways: it may grow in size and fullness; it may grow also in intensity and vigor. Growth of mind in either way is dependent on this attribute of continuousness. Thought thus grows in range and extent and also in fullness and richness, only as the past thought continues, to which a new thought may unite itself and thus add expansion or increase of content. Just so thought increases its power only as the momentum of its past action continues, to which the momentum of a new exertion of thought may be added.

Feeling, in the same way, enlarges its capacity and quickens its sensibility, only as the past feeling continues to swell and soften the heart. He only that has himself felt sorrow can well sympathize with the sorrowing. The larger the experience of sorrow, the larger room is there for sympathy to enter and abide. It is so with all the modes of feeling. Hope expands and enlarges with allowance on its legitimate objects. Love, by taking more and more of its object into its affection, widens its range and extends its sphere. The feelings, likewise, grow warmer and more intense as they are continuously indulged.

Pity and compassion, that have been often moved, respond more quickly and tenderly to new objects. The philanthropist—he who lives a humane and charitable and loving life—has not only a larger but a warmer heart than the ordinary man. The artist, accustomed to contemplate and enjoy objects of art, not only has a larger expanse of sensibility to be touched by beauty, but responds more sensitively to the address of every beautiful form; his admiration is both richer and deeper, both fuller and livelier.

Purpose and endeavor, likewise, grow by virtue of this continuousness of all purpose and endeavor. They grow in breadth and compass. The child can extend its plans, its purposes, only to a very limited range of things; he grows to the large capacity of a man's will and endeavors by adding to the part which he still holds some new field of action. His resolutions and determinations grow also in energy, in strength, as he repeats resolutions and actions, and carries the momentum of the past into the strength of the future.

CHAPTER VI.

THE SELF-CONSCIOUSNESS OF MIND.

§ 26. The mind is conscious of its acts and feelings.

Self-consciousness explained. This proposition indirectly involves the truth that the mind knows and feels that it is itself that acts and feels; that itself is the subject of these changes. It implies, therefore, that each mind is distinct from all other minds, all its actions its own and not another's. Language is sometimes used which seems, if literally interpreted, to teach that our thoughts and feelings and purposes are but the one thought and feeling and purpose of the race or of the universe of rational beings—one continuous vibration, one throb of all being, repeating itself over and over—and so my thought is but the one thought of God or the Infinite One who is the only real existence, and comprises all things in his single nature. But while thought may be common, it is yet distinct and ever the separate act of each separate mind, just as the common bending of the whole wheat field is yet the bending of each individual stalk, and the common flow of the river the distinct movement of every drop, every molecule of water. This is a fundamental element in the nature of self-consciousness, that the mind is distinct from other minds, so that its acts are its own acts and not another's.

§ 27. Self-consciousness immediately respects not the mind itself in its essence, not the conscious subject as distinct from its actions, but more exactly the mind itself as acting in some specific way.

Its immediate object.

There could, indeed, be no self-consciousness except on condition of particular operations or affections of mind, since the mind exists only as acting and its nature imposes upon it the necessity of acting in some specific way. These specific concrete activities of mind are at once the occasions and the proper objects of consciousness. Where there are acts indeed we recognize at once that there is an agent and so distinguish in our thought the act from the doer. Only in thought, however, is there this separation of agent and act. In every specific instance of thinking, feeling, or willing, the mind is of course necessarily present. It is ever the mind thinking, feeling, or willing which is the immediate and proper object of self-consciousness.

§ 28. In self-consciousness there is both knowing and feeling; the mind both knows and feels that its acts are actual and that they are its own. In it, the mind has for itself both light and heat.

In knowing and feeling.

The term *consciousness* comes from a root or stem properly signifying knowledge. But, as is often the case with words, it is used in a broader sense than this, embracing feeling as well as knowledge proper. We as truly feel that the thought we put forth, the feeling we indulge, the purpose we form, exists and is our own as know them to be and to be our own. We accordingly feel displeasure or complacency, a sense of shame or pride in our thoughts and feelings. These thoughts and feelings, besides revealing themselves to our intelligence so that we become cognizant of them, impress themselves on our passive nature so that we feel as well as know them. Indeed, the knowledge, here as elsewhere, presupposes this impression on our passive nature.

In self-consciousness, therefore, we both know our own thoughts, our own feelings, our own purposes and also feel our thoughts, our feelings, our purposes.

The mind's own changes are thus objects to it as really as the changes in the external world. The mind knows and feels its own acts and feelings just as it knows and feels visible and other objects around These two classes of object have sometimes been distinguished from each other; an object without the mind being called *object-object*, and an object within the mind *subject-object*.

Degrees.
§ 29. Self-consciousness still further varies in degree.

The light in which it recognizes the changes in the mind's activity, the warmth in which it feels them, vary from darkness to perfect clearness, from coldness to burning heat. The very finiteness of the human mind involves this. It can no more take cognizance of all its own modifications at any one moment, it can no more feel impressions from all these states at once, than it can know and feel all that passes in the world around it at once.

In self-consciousness the mind is conceived as standing apart from itself and observing its own movements, the volume of its activity rolling on before its eye. It is able only to receive what light and heat come to it from that side which is exposed to its view. It can notice only a part. The great part of its activity lies out of its view. It may be none the less there—actually existing, so that an omniscient eye would discern it; so that what is now concealed from its own sight might come into view by a change of position. Yet in fact a great part of the modifications of its own activity are at the time entirely in the dark; only a small part is in the fullest light that can ever come to it; and the rest are in clouded light shaded in all degrees from undimmed brightness to the confines of perfect darkness.

We recognize, however, certain stages in this varying light and heat. We speak of being fully conscious, clearly or distinctly conscious; we speak of being feebly or indistinctly conscious; and we also speak of being entirely unconscious.

While all mental states and acts are, as a general fact, within the range of consciousness, only a part, a very small part indeed, are at any one time in actual view. An omniscient mind might see them all; the finite human mind cannot. Nor can the human mind distinguish the exact outlines, the form and figure, the exact degree of brightness or intensity in any state, any more than it can mark off the lines which bound the brightened frontage of a rounded column from its receding sides The light and warmth of consciousness are graded down by indistinguishable degrees from the highest to the lowest. They are measured only by comparison and relatively. The human consciousness varies in feebleness and power with native vigor, with bodily health and condition, with advance in age and experience, with growth and culture.

CHAPTER VII.

THE RELATIVENESS OF THE MIND TO ITS OBJECTS.—IDEAS.

Mind known only in its relations to object. § 30. The mind can be known by us only as it exists in relation to its objects.

Its action necessarily respects an object; its passive nature equally implies an object. It is because of this necessary relation of its acts and feelings to some object that it is called *the subject, the conscious subject.*

Of what is called the substance of the mind, it is familiarly said, we know nothing. We can no more perceive any such substance with the inner eye of the soul than with the outer eye of the body. We can know no object in fact but through its attributes; and as the essential nature of the mind is activity, we can have no knowledge of mind except through its action. But action implies an object; and thus it is that only as the mind goes out in its action towards some object, are we able to recognize any form of its activity. Only then can we recognize the fact that the mind exists, and also that it acts in any particular way.

Ideas. § 31. All such objects for the mind are called *Ideas.* We may thus define *idea* in the words of Locke, to be "whatever it is the mind can be employed about."

When we think of the sun, the sun is the object of our thought, and, so far, it is the idea before our minds. When we feel the warmth or the light of the sun, just so far the sun is object to our sensibility and as such object it becomes idea, impressing our passive nature, entering into our souls and occasioning feeling—sensation—in them.

§ 32. Nothing but idea is object for the mind.

Idea sole object to mind.

Certainly when we say that the sun is in our thoughts we do not mean that the real object itself—that the huge, round, burning, material sun—is literally in them, but only that the idea of the sun is in them; we have, in thinking of it, not the actual sun itself in our minds but an idea of the sun. It is only in a kind of figure that we say the sun is in our thought. Such language is allowable; it conveys a truth; it is perhaps the simplest, easiest, most natural way of expressing the fact in words. But the exactness and precision of science impose upon us the necessity of putting a strict construction, and the true construction, into our words and statements.

§ 33. Nothing but mind is subject for idea.

Mind sole object for ideas.

Ideas are certainly not for matter, as we conceive it to be. To the rock, all idea is as nothing. All objects, in so far as they are ideas, are for mind alone.

The conclusion, accordingly, is that mind, as active, and idea are perfect correlatives. Each is exclusively for the other.

§ 34 This relation between mind and idea is twofold;—the mind originates ideas and the mind receives ideas. And conversely, ideas are of mind and for mind.

<small>Twofold relationship of mind and ideas.</small>

The sun in the heavens is but the Creator's idea of the sun realized. The attributes, round, bright, warm, heavy, moving, attracting, and the like, constituted the elements of his idea—his plan, his design,—as he determined to create the sun and make it real. The Creator had an idea of the sun when he made it. The sun is the Creator's idea realized. We, too, have an idea of the sun. Its elements are the attributes which constitute the sun and make it what it is. By apprehending these attributes we obtain an idea of it. The sun, by being thus apprehended, becomes our idea.

The human mind, as well as the divine mind, originates ideas. The machinist forms an idea of a system of machinery. He puts this idea into the matter which he finds formed and shaped for his use—into wood and iron. His idea is in the completed machine, and we, by studying its parts and its adaptations, apprehend his idea; it then becomes our idea of the machine.

The machinist, perhaps, never, with all his study and labor of thought, attained a perfect idea of a perfect machine; but however perfect his idea of it may have been, he never succeeded in realizing it perfectly—in embodying his idea so that it should be just as perfect in the wood and iron. Moreover, our idea of the machine, however perfectly embody ing the idea of the inventor, may not be perfect;—we may not understand it as fully and exactly as the maker. As it respects the human mind thus, neither the idea as originated, as designed and formed by it, nor the idea as apprehended, as received into it, can be regarded as perfect and complete.

The ideas of the Infinite mind, as formed and designed in it, are on the contrary, perfect and complete, and the ideas as apprehended by it and received into it we must conceive to be equally perfect and complete.

The created universe is God's idea; and all specific objects and arrangements in it are his ideas realized. These ideas of his, as thus revealed in his works and ways, are all we know of him. They are, as Cousin, interpreting the doctrine of Plato respecting ideas, justly says, "nothing else than the manner of existence of Eternal Reason."

Each part of God's idea in the universe is a particular idea to us. The human mind can at best apprehend but parts and but partially. Still each part even but dimly apprehended is an idea to us; and what we do not apprehend may be objects to higher or more advanced minds, and all is apprehensible by the Infinite mind. We are accordingly not to conclude that because we have exhausted our power of apprehending any part, any object, any arrangement in God's universe, there may not be other parts of the general idea—other specific ideas—within the reach of other minds.

§ 35. IDEA, as both of mind and for mind, may more fully and exactly be defined to be *any form of mental activity*.

Idea defined.

Mind, as essentially active, exerts itself or goes forth in some form more or less specific. Only as it acts in some such specific way can it act at all, or, since it is by its nature essentially active, even exist; and only as it acts in such form can any other mind know or feel its activity or its existence.

The Creator has put His activity in matter. The creative spirit, as we are taught, brooded over chaos—formless, motionless matter—and shaped its creative ideas in it. Chaos receives from this creative act, form, extension, limitation in outline and shape; receives also moving form, as attractive and repelling force. It, then, as thus formed, comes within the capacity of our minds. We know matter only as thus shaped and formed. Matter in fact is to us only bare receptivity for ideas. Mind, on the contrary, while passive, impressible, and thus capacity for ideas, like matter, is also faculty of ideas, former, originator of ideas. In this lies the grand distinction, so far as we know them, between mind and matter. Mind forms ideas; matter is formed by ideas. Mind originates ideas, matter merely receives ideas. Matter is bare receptivity for ideas; mind is the capacity for ideas, and faculty of ideas.

§ 36. Any expressed idea, that is, any form of mental activity, may be apprehended either as true, or as beautiful, or as good.

Idea threefold.

The sun, as an idea of the Creator, realized and presented to our view so that we can apprehend it or take it into our minds and thus make it our own idea, is to us either true, or beautiful, or good, as we choose or happen to regard it at the time. We may say thus of the sun, in one view, that it is a true thing; in another view, that it is a beautiful thing; and in a third view, that it is a good thing.

These three characteristics of the sun are very unlike such characteristics as round, bright, warm. We can conceive of a sun that should be without one or indeed without any of these characters, although not without some such characters. We can conceive of a sun that should be jagged or angular in its outline; perhaps of a sun that should give no light, or one that should give warmth without light; but we cannot conceive of a sun that should not be either true or false; or that should not be either beautiful or ugly; either good or bad.

Roundness, brightness, warmth, weight or gravitating force, and the other like attributes of the sun, make up and constitute the sun what it is. If we were to suppose one of these attributes to be abstracted from the sun, it would be still a sun,—a different sun, but still a real, existing thing. But the true, the beautiful, and the good do not, by being combined together in any such way, make up and constitute the sun. We cannot abstract one of these attributes and leave any thing behind as real and actually existing. If the sun be not true, it cannot be beautiful or good; it cannot, indeed, be at all. These three, then, the true, the beautiful, and the good, are equally necessary in every object for the mind. It is possible for the mind, however, to confine its view of any object to any one of these three aspects or phases of the object, and thus the object for the time will be in this view true, or beautiful, or good. Indeed it is difficult, if not impossible, for the finite mind of man to take in all these phases fully at once. It yet ever remains possible for it in contemplating any object whatever, to regard any one of these three attributes at pleasure, and to pass from one to the other at pleasure.

§ 37. We have an idea of an object as true when we apprehend it in its essence, in its constitution or its making up; and accordingly regard it as having certain attributes.

Idea as true.

We have an idea of the sun as true, thus, when we apprehend it in its attributes as round, bright, in the heavens, and the like.

Under the general notion of the true, it should be observed here, is comprehended both the true in its stricter sense and the false or the contradictory of the true, and furthermore the imperfectly true, or the mingled true and false.

An object is strictly true, when all its attributes are harmoniously united in it. We apprehend a true sun when we apprehend it as having its essential attributes,—its roundness, brightness, etc.,—all congruously united in it.

An object is false, when it is represented to us with attributes none of which belong to it or if belonging to it in any sense are not congruously united in it. An idea of the sun which should contain in it the attribute of being angular or cold or without gravity or should combine in it incongruous attributes, as being of curved and also of rectilinear figure, is a false idea.

An object is imperfectly true to our thought when some attributes that are true enter into the idea of it together with some that are false; or when the attributes are incongruously put together in it.

§ 38. We have an idea of an object as *As beautiful.* beautiful when we apprehend it as having form, by virtue of which we can receive it into our minds.

We have an idea of the sun thus as beautiful when we apprehend it in its form, by virtue of which we can see its outline, feel its warmth, and the like.

Under the general notion of the beautiful as inclusive of all form, there are comprehended not only the beautiful in the positive and the strict sense, but also the positively ugly and the imperfectly beautiful, or the mingled of the positively beautiful and the ugly.

§ 39. We have an idea of an object as *As good.* good when we apprehend it as having some beneficial effect or result. or a tendency to produce beneficial effect.

Every idea or form of mind must, as the mind is essentially active, regard an end, a result. Every idea of a perfect mind, as of God, must be perfectly good, perfectly beneficial, working only blessing in its proper tendency and result. Every idea of an imperfect mind must have a tendency to a like result as positively good or positively bad or imperfectly good, being mingled of the positively good and the bad.

We have thus under the general notion of the good the three phases, as before, the positively good, the positively bad, and the imperfectly good.

These three great comprehensive ideas, the true, the beautiful, and the good, appertaining alike to every object which can come into our mental experience, are thus distinguished from one another: The true respects the essence—the relation of the attributes as congruously united in it; the beautiful respects the form; the good respects the end in tendency or result.

The three ideas respective objects to the three functions of mind. § 40. The true, the beautiful, and the good are, respectively, objects for the intelligence, the sensibility, and the will.

The true as object thus, is the proper correlative of the intelligence as subject. The true is exclusively for the mind as intelligent, and the intelligence is exclusively conversant with idea as true.

The beautiful as object is, in like manner, the proper correlative of the sensibility. The beautiful is exclusively for the mind as feeling; and the sensibility is exclusively conversant with idea as beautiful.

In the same way, the good is the proper correlative of the will. The good is exclusively for the mind as willing, and the will is exclusively conversant with the good.

The fuller expositions of the meaning of each of these general statements and the proofs of them will be presented in the several books that follow. It is sufficient here to remark that all the facts of mind, all its phenomena, may be distributed into classes either in respect to the mind itself as conscious subject, or in respect to the objects which its activity respects. In the former alternative, classifying the phenomena of mind in respect to the subject, we have the subjective classification,—that which gives the phenomena of the intelligence, those of the sensibility, and those of the will. In the latter alternative, classifying in respect to the object of mental activity, we have the objective classification,—that which gives the phenomena of the true, those of the beautiful, and those of the good. It is obvious to remark that if these long accepted and well established classifications be recognized as correct, they must be exactly coincident; the objective distribution must exactly correspond with the subjective; the objective and the subjective must be exact correlatives.

The difficulties that may seem to beset this recognition of the exact correspondence between these classifications of mental phenomena will be found to pass away as the habits of our thinking become better familiarized with the correspondence, so far at least as the difficulties do not spring from erroneous views of certain facts of mind or representations in language that are inadequate or inexact. It may be that a comparison of the particular results to which the two classifications respectively lead will result in a modification of our views of some of the phenomena of mind. The study of each in the light of the other will certainly help to clearer views and larger knowledge.

CHAPTER VIII.

SYMBOLS OF MIND.

§ 41. We are very prone to think of the mind as we think of something that we have seen; to represent it to our thoughts as if it had form or figure or size. And in speaking of the acts and states of the mind, writers fall very naturally into the use of images and illustrations taken from outward material objects. Thus in its acquisition of knowledge, the soul has been likened to a smooth table of wax on which ideas are written. And Plato speaks of a kind of waxen material in our souls; in one larger, in another smaller; in one purer, in another more impure; in one harder, in another softer. In another place he illustrates the acquisition and the recollection of our ideas under the figure of a pigeon house into which a man gathers the birds he has snared, and in which he keeps them till he has occasion to use them, when he goes and takes out such as he wishes. So we acquire ideas when we catch them and put them away into our pigeon house; we have them there after we have thus acquired them, and when we wish to use them we go and take what we please. But here is the source of mistake and error. As these acquired ideas are flying

Mind Symbolized.

By waxen tablets.

By piece of wax.

By pigeon houses.

about in their cage we mistake one bird for another and take a pigeon instead of a dove, or in our haste and confusion satisfy ourselves with eleven birds when we think we have twelve. In the same way, it has been a very common practice with modern writers to represent the mind as a bundle of faculties: one of perceiving, another of remembering, another of comparing, and the like. In all these comparisons or illustrations there is always the danger of extending the illustration too far. There is no perfectly adequate symbol of the human spirit in the material world. Many objects that we see may represent very vividly something that is true of the mind; but it is ever to be regarded as an imperfect, partial, inadequate representation; and the use of any is to be attended with great care. Subject to this precaution we may advantageously use that marvel of modern skill, the steamship, as a symbol of the greatest wonder of divine wisdom on earth, the human spirit.

By modern steamship.

It will help to present to our thoughts many at least of its attributes more vividly than can be done in mere abstract description. That the mind with all its diversity of faculties is yet one and simple, is imaged in the ship, single with all its diversity of arrangements, capabilities, and outfit. It has hull and rigging; it has sails and steam-enginery; it has keel and rudder; it has ballast and cargo; it has chart and log-book; it can move only in water, but it can make water to furnish it motion; it is impelled by winds and currents from without and is driven also by a power which it carries within. But with all this diversity it is itself one and ever the same.

It well illustrates this attribute of simplicity which characterizes the mind in all the diversity of its states and operations. As it is the same ship in its entireness that moves along, carrying with it sails and engine, rudder and log-book, drives its enginery when it catches the breeze, plies its rudder at the same time that it drifts with the current, so the same one mind in its entireness moves whenever it puts forth any exercise; it thinks while it feels; it remembers while it perceives; it wills while it compares or judges.

The mind's attribute of essential activity is also well imaged in this symbol. The ship afloat is never at rest, not even in the greatest calm. The motion of an object that we can see is the best image we can form, although not an adequate image, of this restless activity of the human spirit. The mind was made to move; to move ever till it reaches its haven and ever towards it. Its activity is not suspended even in the calm of sleep.

The finiteness and dependence of the human spirit is, moreover, well imaged in this symbol. The ship cannot move at all except as it floats and can push its paddles against the very water which bears it. Nor can it even guide its motion but by the aid of this same resistance.

On the other hand, as the ship has within itself its own power of motion, and can drive itself by means of the machinery within itself against wind and current, as well as guide its motion by means equally within itself, so the mind of man is self-active and self-directive while yet dependent on the providence which sustains and which also impels and drives from without the mind and independently of it.

Still further, this symbol images the continuousness of the mind's activity. The course made to-day is different from what it would be but for the course made at the very beginning of the voyage, and the course made each subsequent day will be modified by the course of to-day. If to-day it is driven hither or thither by a breeze of passion, or has been borne away by a current of desire; if it has been put far on its way by an unusual press of its enginery or been guided away far from its proper course by a careless or mistaken management of the rudder; the wind and the current and the whole motive power of to-morrow will find it in a different quarter and move upon it in a different way from what would have been the case had there been different feelings or exertions to-day. And the faithful log-book has all the advances and all the directions, and all the velocities of speed and all the days of tardiness and relaxation of power, traced exactly upon its pages.

And this suggests still another feature of resemblance. The human soul carries its faithful log-book, and it can read the courses and distances of its daily and hourly life. They are all traced there in ineffaceable characters. If in its long voyage it be difficult to look at the records of its earlier movements, or if the few words that record the minuter experiences are lost in the long pages of the great scenes through which it has passed, they are yet all there—thoughts, passions, plans, and purposes,—all there to be found on a closer scrutiny than the hurry of the present hour may perhaps allow. The mind is self-conscious, and its past activity never becomes obliterated from the record of its progress.

BOOK II.

THE SENSIBILITY.

CHAPTER I.

ITS NATURE AND ITS MODIFICATIONS.

§ 42. THE SENSIBILITY IS THE MIND'S CAPACITY OF FEELING AND IMPRESSING.

This department of mental phenomena embraces all that takes place in the passive nature of the mind;—all that it experiences as impressed or affected or in any way moved. Book I., chapter IV. It is so impressed or affected not only by beings and objects without itself; but also by its own acts which react upon it and impress or affect it in various ways. § 19.

§ 43. The mind is said to be in a state of sensibility or of feeling, when sensibility or feeling predominates and characterizes the mental condition.

In all probability the mind is never so completely engrossed with feeling as to intermit entirely its thinking and willing activities. These activities, even in the greatest excitements of feeling are, it may on good grounds be believed, only in relative depression or are less prominently presented to the eye of consciousness.

§ 44. Every mind accordingly stands in this

double relation to other minds : it is impressed by them and it also impresses them.

This first of the three great departments of mental phenomena includes all that the mind experiences in this reciprocal action of mind on mind, whether as receiving or communicating.

The term *sensibility* points more directly to the passive side, to the mind's capacity of receiving idea This department of mind has accordingly been for the most part treated by writers on psychology as if having only a passive side, as if a mere capacity in the stricter sense of that word ; and the corresponding faculty of impressing has been treated as a department of the intelligence under the name of the imagination. But obviously, if the intelligence be recognized as simply a knowing or cognitive function, this classification of the function of impressing is erroneous, for the imagination is not at all a cognitive power. Sensation, it may be remarked here, has also generally been presented as lying in the department of the intelligence, but with as little propriety as the imagination, unless they are regarded as mere conditions or sources or accompaniments of knowledge.

§ 45. We have found §§ 31–34 that ideas are solely of mind and for mind ; and that mind is solely concerned with ideas. Now ideas, so far as they are regarded simply in the light of their being the interchange of minds, are in this aspect, in order to distinguish them from other aspects of ideas, denominated *forms*.

The terms *idea* and *form* are of the same origin.

They are consequently of the closest affinity. They are often properly used interchangeably. But they are not exactly synonymous. When used for exact philosophical statement, form denotes only this one aspect of idea just stated—of idea as interchange of minds or medium between minds.

The sensibility, accordingly, being but the mind communicating or interchanging with other minds, or in an analogous way communing with itself, has to do with form—with form alone.

§ 46. But form as medium between interchanging minds, has two sides or aspects, as idea is both imparted and also received. Form is either form communicating, impressing—*forma formans;* or form communicated, impressed—*forma formata.*

§ 47. In the same way, the imagination, whose sole function is that of form, is either (1) passive or receptive of form—capacity of form; or (2) active or communicative of from-faculty of form. We have thus what is called the *active imagination* and the *passive imagination.*

But inasmuch as the term *sensibility* from its etymology rather points to the passive side, and the term *imagination,* on the contrary, to the active side, it is fitting as it is convenient every way, to employ the former term rather to denote the capacity of the mind as receiving, and the latter term to denote the corresponding faculty of the mind as imparting.

§ 48. The two most generic departments of the phenomena of the sensibility are, accordingly, the *sensibility* proper or the capacity of feeling, and the *imagination* or the faculty of impressing.

The states of the sensibility generally are denominated *feelings*. The products of the imagination are called *forms*.

§ 49. The sensibility is modified in the following general ways:—First, in respect to object, or source; secondly, in respect to purity or simplicity; thirdly, in respect to the intelligence and the will; fourthly, in respect to degree. These modifications give rise to so many different classes of feelings.

§ 50. In respect to object or source, the feelings are distinguished into two classes:

1. Those which flow from the general life of the soul, and attend the exercises of its functions without particular reference to the object, and which may be denoted by the general terms of *pleasure* and *pain*; and

2. Those which are determined by some specific object awakening or producing them. Of this second class there are two species, as the objects are material or purely mental,—*sensations* and *emotions*.

§ 51. In respect to purity or simplicity, the feelings are either *simple* or *complex*. The feelings already named, those of pleasure and pain, as also the sensations and the emotions, are, properly, pure and simple, although entering into other feelings and variously modifying them.

The complex feelings are of two classes, (1) those which simply flow out towards their objects and expend themselves on these objects, denominated *the affections*; and (2) those which reach after their objects to grasp and appropriate them to the mind's own uses, denominated *the desires*.

§ 52. Still further, the feelings enter into all the mind's acts of knowing and willing, and are in their turn variously characterized by them. Insofar as they are so characterized, the feelings are called *sentiments*. They are either *contemplative*, as characterized by the intelligence ; or *practical*, as characterized by the will.

§ 53. In respect to degree, the feelings are modified in indefinite ways, as they vary in intensity from the calmness that borders on apathy or insensibility to the wild passion which characterizes the brute or the madman.

§ 54. In accordance with this general analysis the phenomena of the sensibility will be presented in distinct chapters in the following order, viz. :

>The feelings of Pleasure and of **Pain ;**
>The Sensations ;
>The Emotions ;
>The Affections ;
>The Desires ;
>The Sentiments ;
>**The Passions.**

CHAPTER II.

PLEASURE AND PAIN.

§ 55. PLEASURE and its opposite, PAIN, are the simplest and in a certain sense the most fundamental and pervasive feelings of the soul. They cannot be analyzed nor be defined except indirectly, as by synonymous words, by indicating the occasions on which they are experienced, or through their causes or effects. The best of these definitions, all alike inadequate, is that which represents them to be the mind's experience of good and evil.

Pleasure and Pain.

The term *pleasure*—and an analogous observation is to be made of the term *pain*—is here used in its larger and higher sense as including all experience of good, all happiness, all blessedness, all enjoyment, of whatever kind or degree and from whatever source it may arise.

The entire lawful activity of the soul, we may safely assume, was designed by its Creator to be attended by pleasure; and the entire influence of beings and things around it was equally designed to bring pleasure. The universe was designed, in other words, for good. But evil exists; why and how, it does not belong to us here to inquire. Pleasure and pain are consequently both experienced by man. They each at different times attend the working of the soul's own activities; they each attend the impression made by other agencies on it.

§ 56. They are both, accordingly, distinguishable in our thought from the action which the soul puts forth and from the impression which the soul receives. Only as they enter into such exerted action or such received impression do they enter into the proper substance of the soul to form a part of its permanent active and passive being. The soul does not carry them along in itself except as it carries their causes or sources, or as they modify in some way its conscious activity. The pleasure which I experience from eating an orange, and the pain I experience from the prick of a needle, do not live on in the soul an ever present pleasure and pain, as does every proper act and every proper impression on its passive nature. The memory lives on so far as the pleasure and the pain may have come up into the consciousness; the sensibility bears on with itself the mark of the impression made upon it. But the pleasure and the pain themselves have after a time vanished and never reappear, except with the reappearance of the cause or the source. The imagination brings back or re-creates the act and the impression, and presents its product as colored by the attendant pleasure or pain, but it has no power to reproduce the pleasure or the pain itself.

Distinguishable from mental act.

§ 57. Pleasure and pain vary through all degrees of intensity, from the faintest flush of satisfaction to the brightness of ecstatic joy and from the thinnest cloud of

Degrees of intensity.

discontent to the stormiest violence of grief and agony.

Enters into all mental states. § 58. They intermingle with all the experiences and energies of the mind, penetrating every other affection, investing every movement of the intelligence, and animating or disheartening every activity of the free will.

Sometimes they become prominent and give character to the state of the soul as the leading phenomenon at the time: sometimes, they are lost from the distinct view of consciousness, however much they may yet imperceptibly influence the mind and its action.

They are sometimes momentary and transient, as when a single throb of delight or a single shooting of pain or sorrow is experienced. They sometimes are continuous and lasting, become habitual and mark the permanent character. Thus there are often to be observed moods of gayety and gladness, and also moods of depression and heaviness.

Like all acts and states of mind, as ideas or forms of mental experience, they become objects to the intelligence, when regarded as having certain attributes, for instance as being faint or acute, or as related to their causes or to other objects. They become objects to the sensibility itself as forms for its own contemplation and for affecting other modes of feeling, modifying them in various ways. They become objects also to the free-will as they are allowed or curbed and checked, or resisted and repressed.

§ 59. Pleasure and pain, still further, are variously modified according as the exercise of the mind's activity on which they attend is connected or not with other activities as objects.

Modified by relation of the mental act.

In the first place, there is a natural pleasure attending the simple employment of our mental powers and capacities.

There is a pleasure in feeling, in knowing, in endeavoring. These functions were implanted in us in perfect goodness and wisdom; their legitimate working is the form of the Creator's goodness and wisdom, impressing us as capable of receiving good and enjoying it. This kind of pleasure underlies all other feelings and pervades them all. It does not often, in its lighter degrees at least, come up into distinct consciousness.

The less intense but more permanent and constant form of this pleasure is exemplified in the general cheerfulness of spirit which attends our active life; as the opposite feeling is instanced in the *ennui* of inaction. It is also exemplified in the bright serenity and contentedness that wait on the sound and healthy action of the divers bodily functions; and its opposite in the sadness and heaviness which continued ill-health naturally occasions.

It rises with increased activity. Vivid impressions moving the sensibility to an unwonted warmth and glow, vigorous thought, and energetic resolve, carry this kind of pleasure to the degree of rapture and ecstatic delight. In like manner, the physical functions may play their part so vigorously as to occasion great buoyancy of spirits and gleefulness, while obstructions to their proper exercise induce heaviness and gloom.

§ 60. In the next place, this pleasure that waits on the legitimate exercise of the mind's activity in any specific way is enhanced and characterized when this specific activity is in harmony with the proper play of the other functions of the mind.

By relations to other functions.

Thinking is more pleasurable when the sense which introduces the objects of thought is pleasantly and properly engaged and when the purposes and dispositions of the free-will are furthered and strengthened by the thought. On the other hand thinking is less pleasurable, and may be actually painful when thought is pushed on with an offended sense and an opposed unwillingness or indifference. In the same way the pleasures of sense are enhanced when the curiosity of the intelligence is fed and the purposes of the will are furthered; and the action of the will gives likewise higher pleasure when it moves in pleased feeling and in clear light.

§ 61. In the third place, this pleasure naturally belonging to the mind's activity is enhanced or impaired by its moving in harmony or in disharmony with the bodily functions.

By relations to the body.

A sound and active mind finds its proper pleasure and satisfaction furthered or hindered according as the bodily health is sound and vigorous or otherwise. Still more, any species of mental activity that properly engages any function of the body or works itself out through it, is helped on to a higher pleas-are when the action finds free and harmonious expression through the bodily organ. Vigorous thinking through a brain that is pervaded with diseased or partially paralyzed nerves is so far made hard, wearisome, or painful. If on the other hand, such thinking puts the brain and the whole sympathizing nervous organism into legitimate play, the natural pleasure of the exertion is greatly increased by this harmony between the mind and its physical organs.

§ 62. Still further, this pleasure of normal activity is modified as it is in harmony or disharmony with the action of other beings and the flow of providential events.

<small>By relations to outer objects.</small>

Just so far as these influences that come in upon the mind from without, or that are encountered by the mind as it puts forth its activity, oppose and obstruct its actions, its proper pleasure is marred.

Thus we find the ordinance of the Creator to be ever peremptory in requiring of us, as we desire happiness, first, a full activity of all our powers; secondly, a full harmony between all the several functions of our spiritual being; thirdly, a free reciprocity and concord between our mental exercises and our bodily functions and conditions; and, fourthly, a harmonizing of all our actions with the flow of God's providence about us.

CHAPTER III.

THE SENSATIONS.

Sensation defined. § 63. A SENSATION is an affection of the sensibility by or through the bodily organism.

From the mysterious union of the human spirit with a material frame-work, there comes to it divers impressions which are readily distinguishable from all other feelings. There are pleasures and there are pains which come into our souls through our bodies,—which are, properly and characteristically, physical and corporeal. These pleasures and pains are variously modified; so that we not only distinguish them generally from other feelings, but also distinguish varieties among them that differ in kind from one another. The feeling that arises from the laceration of a nerve differs from that which springs from hearing a harsh and discordant sound not only in degree, but also in kind.

Has its seat in the mind. § 64. All sensation although having its seat in the mysterious union of mind with its bodily covering, is yet to be referred to the mind alone.

The body is affected, it is true, in all sensation; but it is not the body that feels. It is not the eye or the ear that feels the light or the sound in the sensations of sight and hearing, not the nerves that as matter convey the light and sound to and through these organs, not the brain in which these nerves finally terminate, and, so to speak, deliver what they have received; it is the mind, the soul itself only, that feels in sensation. Matter has no sense; not even organized matter has any sense in any proper meaning of that term.

Medium of sensation. § 65. The medium of sensation in the human body is seated in the nervous organism, embracing the nerves proper, the spinal marrow, the brain, and the other collections of nervous matter, placed in various parts of the body. This system of nervous organization is collectively called the *sensorium* or *sensory*.

It is through this nervous organism or sensory that the mind receives and imparts whatever is received or imparted by it in its communications with external beings or objects. Whether the contact of mind with matter is limited to a point or is extended over more or less of the sensory, whether, indeed, it is not absurd to speak of any localizing of the spirit, are questions which lie in darkness too profound for science at its present stage to determine.

Two classes of nerves. § 66. The nerves are of two classes: (1) *sensitive nerves;* (2) *motor nerves.* The sensitive nerves, called also afferent nerves, receive and transmit to the mind; the motor nerves, called also efferent nerves, impart from the mind. Through the former, the passive nature of the mind is affected; through the latter, its active nature goes forth to impress other beings or objects. Accordingly the motor nerves are not immediately acted upon by external objects. In order to move them, the external object must first irritate a sensitive nerve, which conveys the impression to the mind, and then from the mind or at least from the mind's more immediate organ, the brain, goes forth the force which excites the motor nerves. This is, at all events, the general law. It is thought, however, by some physiologists that the motor nerves and the sensitive nerves when in juxtaposition may reciprocally influence each other. Moreover, if a galvanic current be made to pass along the nervous fibre in the outward direction from any point even for the slightest distance, the muscles to which the nerve is attached are at once contracted. No such muscular contraction arises if the current be transmitted in the reverse direction from without inward. However the mind may be remotely affected by these reciprocal influences of one part of the bodily system upon another, these influences and their effects are in themselves merely physiological, and call for our consideration only as they either come finally to impress the mind and awaken feeling in it or are themselves first originated by it.

Same sensation from whatever part of the nerve impressed. § 67. It is a most noticeable fact that to whatever part of a given nerve the stimulant is applied,—to its extremity where it terminates in the muscle, or to its root in the brain, or to any intermediate point,—the sensation felt by the mind is the same.

Thus is explained the fact that a blow or a pressure on the eye, even when the eyelids are closed or in perfect darkness, occasions the sensations of sight, and the mind sees forms and colors that exactly resemble those produced by visible objects from without. In the same way chemical irritants, as narcotics, the circulation of impure blood, various diseases, as fever and hysterics, and also heat and electricity, have been observed to affect the nerves within the body and cause them to transmit the same sensations as if the nerves had received the impressions on the surface of the external organ. Sensations of heat and cold, and itching and creeping sensations really produced from within, as from the stomach or other internal organ, often seem to the mind to be on the surface of the body. The mind itself acting on the motor muscles seems, in some morbid or highly exalted states of the sensibility, to be able to impress the sensitive nerves and so to bring to itself sights and sounds exactly answering to those afforded by external objects.

Apparitions accounted for. In the light of this remarkable fact we are enabled to account for many of the phenomena of apparitions and mysterious sounds where there can be no room for supposing any deception or imposture. The sensations felt in the mind in those cases are real; for the nerves are in the state in which external forms and sounds are experienced, and sensation is only the mind's state as affected by a state of a nerve or some part of the nervous organism. The mind actually feels that affection of the nerve; the sensation is therefore real. But instead of an external object impressing the nerve, it is the action of the mind itself or the internal force that has affected the nerve; yet this affection, so far as the mind is concerned, is precisely similar to that which would have been produced by an external object. There is, it should be observed here, a remarkable power of habit which is often operative in these phenomena. This phenomenon is instanced in the case of an old snuff taker. He had been seized with epilepsy. In order to restore him, his nose was tickled with a feather. The right thumb and fore finger were immediately contracted as if to take a pinch of snuff.

<small>Four classes of sensations.</small> § 68. The sensations may more conveniently be considered under four general classes.

1. The sensation of simple *bodily pleasure and pain;*

2. The sensations that may be referred to the general state of the body as a living system—those of the general *vital sense;*

3. The sensations referable to certain general systems of organs, of which the most important are those of the muscles and of the skin—the sensations of the muscular and the cuticular sense—those of the *organic sense;*

4. The sensations of special organs or those of *special sense.*

<small>1. Simple bodily pleasure and pain.</small> § 69. 1. There is a pleasure attending the legitimate use by the mind of the body as its organ both in receiving and imparting.

There is a pleasure in seeing, in hearing, even when the objects themselves are offensive and the sounds are discordant;—a pleasure which may overcome the displeasure from the sense of the objects themselves, or may only suffice to modify and lessen that displeasure. There is pain when this use of its organ by the mind is impeded. This fundamental pleasure, attending on the legitimate use of the body, and this fundamental pain attending on its abuse or impeded use, pervade all the other more specific sensations, and enhance or otherwise modify them.

To this class belong those sensations which spring from within—from the condition of the body itself irrespectively of any outward cause or influence; such as the sense of health and physical soundness; those incident to youth and to age; those of languor and heaviness; those attending disease or injury of any kind to the body.

It is to be remarked that the sensation, especially the sensation of pain, becomes more or less intense according as the mind yields itself in its general activity more or less to the affection of the sensory. A pain, thus, which if the mind is strongly drawn off in another direction remains unnoticed, becomes almost intolerable if the mind is surrendered to the affection. The soldier in the excitement of battle remains unconscious even of severe wounds till after the struggle is over. An oppressive pain in the head is sometimes lost sight of in vigorous study, or other earnest endeavor. It is in this fact that we find the principal basis of culture for the physical sensibility. It is not so much that the nerves themselves become more sensitive; indeed, they may become from use more dull of impression. But the mind learns how to receive to itself more fully the affection of the sensory. The blind man thus learns to receive more perfectly the sensations of touch, and to interpret them more easily and more accurately, because of the training of his mind in relation to this sense, rather than by any increase in the impressibility of the sense itself.

§ 70. 2. The sensations of the second class, those of the general vital sense, embrace the sensations produced by the atmosphere or other surroundings of the body, such as the feelings of heat and cold, of exhilaration or of depression arising from the state of the air, and the like.

2. Of the vital sense.

These sensations differ from the preceding class in being referable to external causes or influences. They differ from the two following classes in this, that they are not referable to any particular part of the bodily system.

§ 71. 3. The sensations of the third class, those of the general organic sense, are such as are referable to some affection from without of one or other of the organic systems in the bodily structure. They are such as affections of the surface by some foreign body as in titillation; those of heat and cold from contact of external objects or felt nearness to them and generally what may be called affections of the cuticular sense; those of hardness and softness; of weight and pressure; of resistance, and generally

3. Of the organic sense.

the affections of what has been called the muscular sense.

4. Of special sense.

§ 72. 4. The sensations of the fourth class, those of special sense, embrace those to which a special organism is appropriated in the bodily economy. Of these special organisms there are five in number to be recognized, viz.: those of Touch, Taste, Smell, Hearing, and Sight.

The sense of Touch.

§ 73. I. The sense of touch has its seat in the extremity of the nerves terminating in the skin, particularly at the tips of the fingers, or the tongue and the lips.

At these points the sensory *papillæ*—the elevations of the surface of the skin in which the nervous fibres terminate—are most numerous. It is only by the greater number of nerves terminating at these points, and consequently the greater sensibility of these parts, that the special sense of touch can be distinguished from the general organic sense residing in the skin; and some physiologists have therefore not without some reason identified the sense of touch as a part of this general organic sense. The relative acuteness of the sensibility to external objects in different parts of the body has been determined by ascertaining the smallest distance of separation at which the points of a pair of dividers could be distinctly felt. It is found that at the point of the tongue this distance is about half a line or one twenty-fourth part of an inch; at the end of the third finger, one line; on the lips, two lines; end of the nose, three lines; the cheek, the palm of the hand, and end of great toe, five lines; the knee and back of the foot, eighteen lines; middle of the back, of the arm, and of the thigh, thirty lines.

If the touch as a special sense be distinguished from the general organic sense, it can be impressed by matter only through its mechanical and special properties. It can be excited only by actual contact of the external body with the tactual organ. The general organic sense, on the other hand, is excited by heat radiated from remote bodies.

The touch is the earliest of the special senses to be excited. It is closely connected with the organic, as the muscular and cuticular senses, on the one side, and with the special sense of the taste on the other. Its sensations mingle with the other sensations so freely and so thoroughly that it is often difficult to determine to which the mental affection is most to be attributed. This mental affection is thus modified in indefinite modes and degrees.

Of taste. § 74. II. The sense of taste has its seat in the tongue and in the soft palate with the adjacent parts.

These parts are covered with *papillæ* in which the gustatory nerves terminate. The sense is excited by actual contact with the organ of the external object in a liquid or gaseous form and near the temperature of the body. It is in close connection with the organic senses, and with the special senses of touch and smell. It is affected only by the chemical properties of bodies.

THE SENSATIONS 71

Of Smell. § 75. III. The sense of smell has its seat in the inner and upper part of the nose. It is excited by odoriferous particles of extreme minuteness that are borne to the organ by currents of air, or as some suppose by vibrations of odoriferous ether in analogy to the sensations of sight from luminiferous ether.

If the breath is drawn through the mouth, there is no sense of smell. Its sensibility is weakened or destroyed if the inner lining of the nose is too dry or too humid, as is familiarly experienced in colds. Its affections, except when extremely vivid, do not engage the mind unless its attention is turned towards them. They are easily overpowered by the other sensations and in other mental operations. They seldom engross our thoughts. The sense varies greatly in susceptibility in different persons; the cases of its being entirely wanting are not infrequent. Some persons are extremely sensitive to certain odors to which others are indifferent. Some odors are agreeable to some persons, which are offensive to others. It is closely associated with the special senses of taste and of touch and also with the general organic sense. It is excited only by the chemical properties of bodies.

Of Hearing § 76. IV. The sense of hearing has the ear for its special organ.

It is ordinarily excited simply by the mechanical effect of waves of air put in motion through some elastic or sounding body. These waves are first received in the external expansion of the ear, and are from that transmitted by a marvelous complicated apparatus to the auditory nerve. If the waves of sound strike upon the ear at unequal intervals only a noise is heard; if the waves are equal and move at equal intervals, the sensation of musical sound is produced.

As the seat of sensation is in the auditory nerve, which lies deep within the head, any agitation of the extended apparatus for hearing is subject to divers affections within the outward ear which may occasion shocks to the nerve and thus give the sensation of sound. Thus is explained the familiar fact that both noises and ringing sounds are frequently heard when there is no external sounding body to produce them, but only some muscular disturbances arising from within—from disease, from local inflammations, from narcotics or the like. The circulation of the blood thus often occasions sensations of rumbling or roaring when the nerves are morbidly tender and irritable. The same electric shock, indeed, transmitted through the body has been known to give at the same time sensations of sound, of light, of smell, of taste, and of touch. In this case the nerve itself seems to be immediately acted on by the electric fluid.

The sense of hearing seems more isolated than the three preceding senses. It is excited, if we except electric action upon it, and perhaps mental irritation through the motor nerves or disease in the auditory nerve itself, only by mechanical impulse.

Of Sight. § 77. The sense of sight has for its special organ, the eye.

The impressions of the visible body, or, more

scientifically speaking, the undulations of light from the body, are transmitted through the outer parts of the eye to the retina or expansion of the optic nerve, and then through it to the brain. An excitement of this nerve in any part of it may occasion the sensations of sight. Accordingly, so called spectral illusions may be caused by blows, by diseased condition of the humors of the eye, by nervous irritants, by electric excitement, and also by purely mental agitation reaching the optic nerve through the nerves of motion. It is supposable therefore that there may be true vision even when the outer eye is destroyed, or the retina paralyzed, provided that the optic nerve at any point is sound and can be reached in any way by any irritant as the electric fluid or the motor nerves acted upon by the mind.

The sense of sight, like that of hearing, is more isolated than the three other senses. It is excited, if we except affections from within the organ and from mechanical violence, only by the undulations of light. It is subject consequently so far to the physical laws of light, as, for example, those which govern its velocity, direction, intensity, decomposition, and the like.

CHAPTER IV.

THE EMOTIONS.

Emotion defined.
§ 78. An Emotion is an affection of the sensibility immediately by an idea, that is, by some form of mental activity.

An emotion differs, accordingly, from a sensation in this, that the latter is a feeling excited by the bodily organism,—the soul being impressed by the idea mediately through the material body;—while the former is directly excited by purely mental or spiritual acts or states,—the soul being immediately impressed by the idea.

It ever respects immediately the form of the idea, not its essence nor its end or result. Thus when it is the truth of the idea, the idea as true, which addresses the sensibility, the impression on the sensibility, or the feeling awakened is logically before the actual apprehension of the truth by the intelligence, as something to be known. It may indeed be contemplated for a longer or shorter time, before the intelligence comes to recognize it, to accept it as true, to know it. It is simply the truth as form to be impressed on the mind that the sensibility is concerned with. The emotion may indeed continue after the truth is accepted as true and is perfectly known; but the emotion itself is a feeling, not a knowing state of the mind, and precedes the knowing as its ground or condition.

The emotion may be awakened by an idea originated by one's own mind or presented by another mind. The idea in the former case has been called *subject-object*, to distinguish it from the idea in the latter case which has been called *object-object*.

Three classes of emotion. § 79. We have recognized three general classes of ideas, the true, the beautiful, the good. There are accordingly three general classes of emotions:—

1. The emotion awakened by the idea as true impressing the sensibility;
2. The emotion awakened by the idea as beautiful; and
3. The emotion awakened by the idea as good.

The emotion, it should be observed, is ever pervaded and accordingly characterized, more or less, by the fundamental pleasure which attends all legitimate mental action. Where the idea that awakens the sensibility is subject-object, there are really two combined sources of this pleasure, which, consequently, other things being equal, is so much enhanced.

Thei name. § 80. In analogy with the designation *bodily sense*, a phrase derived from the source of the impression on the sensibility, the three classes of emotions may be respectively called the *Intellectual Sense*, as the intellect is the source of the idea as true; the *Æsthetic Sense*, as the æsthetic relation is the source of the beautiful;

and the *Moral Sense*, as the moral nature is the source of the good.

The emotions of each class are variously modified.

Intellectual sense
§ 81. The emotion of the true or the intellectual sense, is awakened by simple truth.

There is a native aptness in the mind to be impressed by truth.

The impression may be lost in the knowing act that follows and be unnoticed; but it is still real; and often the feeling of truth is strong and permanent and characterizes the whole mental state. Falsehood, on the other hand, while it impresses as bearing the general form of truth, yet is naturally attended with a feeling of displeasure.

Emotion by originality.
A truth originated in one's own mind, and especially if wrought out by great exertion and with much difficulty, impresses the sensibility with a peculiar force. When the mode of weighing King Hiero's crown flashed upon the mind of Archimedes, he leaped from his bath in an ecstasy of joy at the discovery.

By interest.
The emotion, moreover, is more or less complete and perfect, according as the truth itself is more or less closely related to the mind, more or less coincides with its tastes, or its interests.

By object.
The emotion varies too with the character of the truth or object.

Surprise.
If it be new, the emotion takes the particular form of *Surprise*.

If it be new or strange and also imper-

The ludicrous. fect, that is, if there be striking incongruity in its parts or elements, the emotion is that known as the emotion of the *ludicrous*. When the Irish gentleman gravely remarks on the unhappy condition of childlessness, that he has observed it frequently to be hereditary, the incongruity of the attribute of *hereditary* with that of *childlessness*, excites our laughter. A like occasion of the feeling was furnished when a judge of the King's Bench remarked in a will case that 'it was very clear that the testator intended to keep a life interest in the estate to himself.'

Wit. As an idea is a form of mental activity, if it bears the character of a quick and vigorous mind, the emotion which it awakens becomes more vivid. The feeling is strong thus when awakened by quick wit, as in repartee. As when once a grave and eloquent lawyer, in an important argument, had observed in emphatic terms, " you cannot split hairs in such a question," and paused for a moment, looking sternly and triumphantly down upon his opponent who sat before him playing with a penknife, and when the latter, not at all disconcerted, instantly pulled a hair from his bristly head and held it up fairly split, to the great amusement of the court and jury, the quick wit of the advocate turned the tide as he added : " I said a hair, sir, not a bristle."

§ 82. The æsthetic sense is awakened Æsthetic sense. by the form of an idea. This form when most perfect is none other than beauty ; beauty is perfect form. Whatever is beautiful moves us by a power of its own. There is a native aptitude in the mind to be impressed, and if the mind is in sound condition, and free to be impressed, and the idea that impresses it is itself perfect in its form so as suitably and freely to impress, the emotion caused by the impression is attended with that fundamental pleasure which waits on all legitimate action.

Power of beauty.

As in the case of the intellectual sense, the emotion is variously modified by the object which impresses it, and is characterized by that. The emotion of the beautiful varies accordingly with the nature of the form which awakens it. It varies as the several elements of beauty or of form vary; as the idea, the matter in which it is revealed, or the rendering of the idea in the matter varies. It varies with the relative predominance of either of these elements. Thus we have *the sublime* when the idea predominates over the matter; *the proper beautiful*, when the idea and the matter are just commensurate with each other; and *the pretty* and *the comic*, when the matter outmeasures the idea.

Emotion varies with the form.

An instance of that species of the sublime in which the grandeur of the conception is expressed in the simplest speech is that so often cited from the time of the Greek critic, Longinus, from Genesis: "and God said, let there be light, and there was light." The few simple words reveal a power which transcends all thought as well as all expression, and which, as the contemplation strives to reach and to grasp it, only becomes the more grand and incomprehensible.

The sublime.

This is the sublime of simple greatness. The sublime of intensity is illustrated in the few quiet words in which Shakespeare makes Brutus intimate the unutterable, inconceivable grief and sorrow that the death of Portia was giving him: "No man bears sorrow better; Portia is dead." There is no limit to the imaginable depth and intensity of anguish here intimated. The longer and closer it is impressed on the sensibility the more is the imagination stirred and elevated. The sublime in nature is exemplified in vastness of extent, as in the ocean, whose bounds the imagination strives in vain to compass, or in the starry heavens shutting in within its limitless expanse countless worlds and systems of suns; and in vastness of power as in the fury of a storm, the crash of the thunderbolt, or the throes of a volcano. The sublime ever exalts, expands, lifts the finite soul, as it were, from its footing and throws it from its balance; it accordingly agitates and disturbs the soul in contemplating.

The proper beautiful. The beautiful, on the other hand, ever tranquillizes and harmonizes. The idea here is in just equipoise with the matter in which it is revealed, and the contemplation sympathizes and is itself quiet and at peace. The ocean is sublime, especially if lashed to a furious display of its power by a tempest; the placid lake is beautiful, as all forces are in equilibrium. The sky is sublime when it reveals to us the infinite extent of creation and the power that made and that upholds and rules it; it is beautiful as its perfect order is regarded, and all the ideas of the creative mind appear wrought out in perfect harmony.

The comic. The *comic,* in which the idea is overpowered or outspanned by the outward form or matter in which it is revealed, in which this outward form arrests the attention to the repression of the idea, neither awes like the sublime, nor simply tranquillizes and leaves in perfect satisfaction like the beautiful. Its proper tendency is to agitate, but the agitation is the surface-ripple, not the deep-rolling billow; it culminates in "laughter holding both its sides," as its natural expression.

The pretty. The *pretty* is the comic bordering close on the beautiful; the idea is in defect, while the outward form satisfies. We distinguish thus a pretty face, a pretty figure, from one that is beautiful, as in the latter idea character is a leading element; in the former it is wanting. Where the comic borders on the sublime, we have the effect expressed in convulsive laughter.

The intellectually ludicrous is distinguished from the æsthetically ludicrous by this; — that while they both address the sensibility, the former consists in a disharmony between the elements of the idea, as in absurdities or surprises of thought; while the latter consists in a disharmony between the essence and the form of an idea, or between the elements of the form itself. Thus a coxcomb is æsthetically ludicrous as he puts on an exterior which there is no character within to warrant. So all pomposity, strut, pretension, is ridiculous. The clown, the buffoon, the harlequin, is æsthetically ludicrous as his dress, his words, his acts, have some show of reason but come short of it; and the unreason predominates. The humorist, the wag, the punster, are intellectually ludicrous; the sallies and the coruscations of their wit give the pleasure; the pedant, the mannerist, the prig, as also affectation, foppery, airs, amuse us from their lack of sense or of capacity as compared with the outward appearance which they assume or affect. The proper comic—the *unreason*—may exist alike in the sallies of the witty and in the blunders of the stupid. The stronger effect of the former, is attributable to the mental activity which has originated them.

§ 83. The moral sense is awakened by the good or its opposite in their several varieties. The moral sense thus respects an idea or form of mental activity in relation to its end or aim. As there are three different views which we may take of an action so far as it is moral, each necessarily implying the other,—the aim or intention, the end intended, and the movement of the aim towards the intended end,—the moral sense is awakened indifferently by either; we are equally and similarly stirred by an action whether we regard the love that prompted it, the happy result which it accomplishes, or the rectitude of the action itself by which the love secures the good. The emotion, it is true, is in some respects modified in the several cases. But it is in its more essential

Sidenote: The moral sense. Respects the end of an action.

character the same. The good action calls forth our approval, our commendation, our complacency whether we regard its motive, its result, or its proper character as right.

In the same way, the action morally bad alike calls forth our disapproval, our condemnation, our displeasure, whether we regard more prominently the hatred that prompted it, the harm which it produced, or the unrighteousness in which it was practiced.

The moral sense is similarly affected whether the action which impresses it, is our own or another's; but the remorse which we suffer from the sense of our own wrong-doing, is generally with imperfect men, affected by their own selfish dispositions; it is less easily awakened, and yet when awakened, is more poignant and bitter than the indignation which is provoked by the wrong-doing of others.

The moral sense, the sense of right or of goodness, is, like the sense of truth and the sense of beauty, a necessary part of every rational nature. Like them, too, it ever answers truly to the object, right or wrong action, good or bad conduct, beneficence or injury. It lies at the foundation of all morality, as the sense of truth at the foundation of all knowledge, and the sense of beauty at the foundation of all art; without it, morality must necessarily be a matter of indifference. It may be blunted, like the sense of sight, by disuse. It may be, like the bodily senses, more acute and quick in some than in others. It is susceptible, like them, of culture; especially may the sensibility be trained to interpret more readily and accurately the character of the object, and thus become more quick to detect and feel the good and the wrong that falls in its way. It may be in particular cases more or less predominant, or more or less subject to other mental occupation; be seemingly lost in such occupation, or be aided and enhanced by it. Habit, taste, interest, may hinder or further its presence in the soul. But it is ever there, a present element, ever capable of being roused whenever its object, right or wrong, is before it, ever alike true in its response to it. The moral indifference or insensibility of a man, or of a community to a wrong, if it be not regarded merely in relative degree, is to be ascribed to the predominating occupancy of the mind by other pursuits, other passions, other interests, which prevents the immediate presentation of the wrong to the moral sense, rather than to the entire want of moral sensibility or to the fallaciousness of the sense itself. If a man, with organs complete, does not see, or if seeing, mistakes black for white, it is not because the sense of sight is wanting or because the sense distorts the image of the object, but because his mind is otherwise absorbed or because he confounds what he sees with other impressions, and so himself mistakes a true vision. That this may be, can be easily supposed if we compare the different impressions made on two different persons of perhaps like temperament and habit, who may be looking out from a rocky shore, watching the sails that are tossing out on the ocean when raging madly in a tempest; one free to admire the awful sublimity of the scene and resigning his whole soul to the impressions of his æsthetic sense, while the other is burdened and oppressed by anxiety for the safety of some precious one whose fate is at the mercy of the merciless elements. Every succeeding moment intensifies the rising rapture of the one, and equally the swelling anguish of the other. The same scene moves each alike; the same sensitive nature belongs to each; but the object impresses a nerve of pleasure in the one, a nerve of agony in the other.

CHAPTER V.

THE AFFECTIONS.

Affections defined.
§ 84. THE AFFECTIONS are feelings which go out and expend themselves on their objects.

The classes of feelings hitherto described are properly subjective. The first class, simple pleasure and pain, are purely so; they do not directly even suggest any exterior object, inasmuch as they are the immediate attendants on the mind's own acts and states. The sensations and emotions only imply and presuppose their objects, and do not actually look out upon them. We are enabled by them alone to distinguish ourselves as subjects of impression from the objects which impress us, but only as passively affected by them, not as acting towards them. Much more than bare implication of objects is attributable to the affections; they are turned directly towards their objects; they fasten immediately on them, and in a true sense go out to them, and lose themselves in them.

Founded in sympathy.
The radical principle or element in the affections is that constitution of the mind by which it, of its own nature, symthizes with all surrounding objects that are akin to itself—to its nature, its habits, its tastes, its pursuits, its interests. It is the principle of kind—

kindliness—attesting the common fatherhood of ourselves and of all objects with which we have to do. Its opposite is antipathy, which feeling is awakened by objects that do not at the time appear in this relationship of being objects akin to our natures or dispositions.

Classed. § 85. The affections are generically classed as *love* and *hate*; love being a synonym of sympathy, and hate of antipathy.

It is to be remarked that as the affections tend to pass out into the purposes and endeavors, love and hate very commonly denote actual benevolence and malevolence—practical benefiting and harming They are, however, properly and primitively, affections; and, in so far as they are affections, the soul is predominantly passive in them, the activity involved in the exercise of them being rather simply that of the will in holding the soul to their objects to be fully impressed by them through that sovereign control which the will possesses over all the functions of the mind.

Varieties. § 86. As habitual dispositions the affections are known by such names as *complaisance, kind-heartedness, charitableness, good-nature;* or *ill-nature, uncharitableness, unfriendliness, misanthropy,* and the like, making divers modifications.

Characterized by object. § 87. The affections are variously modified also by the character of their objects. We have thus love of kindred, love of friends, love of country, love of God, love of animals also, and of inanimate objects, as likewise of places, influences, principles, forms, pursuits.

§ 88. There is, moreover, a class of the affections which are properly responsive in their character. They are called by the generic name of *resentments*.

<small>Resentments.</small>

To this class belong such affections as *gratitude*, that responds to kindness from others ; *anger*, responsive to unkindness ; *forgiveness*, responsive to reparation of unkindness, and the contrary disposition of *inexorableness, relentlessness. Remorse*, when it passes to self-reproach, and so to speak, self-hate, if anything more than the pain that naturally waits on conscious ill-doing, is to be classed among the resentments, as it is responsive to the feeling of guilt ; so also is its opposite, *self-approval*, or *self-complacency*, which in its several modifications in degree becomes self-conceit, vanity, pride, or humility, modesty, and the like.

CHAPTER VI.

THE DESIRES.

Desires defined.
§ 87. THE DESIRES are feelings which not only go out towards their objects but also reach after them to grasp and appropriate them.

We may love without desire; as love may rest upon a present object while desire regards an object that is not present or in possession. Desire implies a want. It is seated in the propensity which necessarily belongs to all life to reach its proper end. An active nature seeks to act, feels impelled to act by a want in its very nature. The desires are all propensities springing from wants or founded in them.

Classified.
§ 90. Desires are instinctive, normal, healthy; or are acquired, irregular, morbid.

It is possible to acquire desires that do not properly belong to our nature. We observe, thus, for example, desires for certain articles of food or drink for which there is no natural taste or inclination, as for tobacco, for alcoholic drinks, for many artificially prepared dishes. It is so throughout the entire realm of the desires. Such acquired desires may have the strength and permanence of a natural appetite, conforming the whole mind and body to them; they may even become hereditary and be propagated.

Desires are irregular either in degree, becoming immoderate and excessive, or in respect to their objects, being turned on that which is unworthy or that which belongs to another. The natural desire for ease and security, thus, must often be stifled when other dearer interests of the soul are in jeopardy.

Desires, still further, sometimes spring out of morbid conditions of the mind or body, and have the strength and importunity of natural instincts.

Desire and aversion. § 91. The desires distribute themselves at once into the two great classes of desires to be in possession of their objects, and desires to be free from them. This latter class are termed *aversions*.

Sub-divisions. § 92. The desires and aversions, further, may be subdivided into as many different varieties as there are different capabilities to be recognized in the soul.

Self-love. In the first rank is to be enumerated the fundamental desire which belongs to every living nature, that its own being be continued; that it be maintained in the highest perfection, and bring in the most and highest good. This is *self-love*, the irregular form of which is *selfishness*.

As every desire directly and necessarily respects a good as its immediate object, and as whatever is good must so far be desirable, life and well-being are natural and legitimate objects of desire. The soul revolts from the thought of ceasing to be—from the thought of annihilation. It is one of the strongest and most deeply-seated desires of a healthy soul that its existence be continued. What will not a man give in exchange for his life? The desire for immortality is accordingly a native, and hence a worthy principle in man.

This fundamental desire for continued existence carries with it the desire that that existence should be as complete and perfect as is possible; that all the capabilities and functions of the soul be in most perfect condition, and that the comprehensive good designed by the Creator in giving and fashioning life be secured in the highest possible degree.

This general desire for life, and for a perfected life in well-being, comprehends desires for the perfection and well-being of each particular capability. We have thus the divers classes of desires belonging to these divers powers and functions and capacities for good.

§ 93. We have, in the first place, the desires belonging to the physical constitution, more commonly known as the *appetites*.

Appetites.

This class of desires respect the continuance and well-being of the individual animal life, as those for food, and for drink—hunger and thirst; the desire, also, for muscular exertion and for rest and sleep. The appetites are all subject to periods. They legitimately arise when the bodily welfare bids When gratified they cease their craving and disappear till the health and well-being of the body reawaken them. The social nature of man is so connected with these animal desires, that they are best gratified only in society and companionship. The taking of food thus tends among all classes of men to become social. There is also to be enumerated here as a proper social desire, that for wedlock, which seeks the continued existence not of the individual, but of the race.

§ 94. We have, in the next place, the proper *rational desires*, which are seated in the soul itself, and which seek the continuance and perfection of its several capabilities and functions.

<small>Rational desires.</small>

Of the rational desires is to be enumerated, First, the *desire of freedom*. The attribute of free personality is the characteristic attribute of man, distinguishing him from the beast. The desire of freedom, in the most comprehensive sense of the word, so far as is allowed in the relations of man, is one of the strongest and most unquenchable principles of the human soul. That it is thus in its very nature of such strength, only indicates the necessity of fostering and also of guiding it. It may become irregular; it may become morbid; it is capable of driving to the worst excesses. Hence the necessity for guarding and regulating it.

<small>Of freedom.</small>

Secondly, *the desire of power* is involved in the general desire for the perfected well-being of the soul. Power inheres in its very being as essentially an active nature; and as it is subject to growth, the increase of power is a natural desire. Ambition is thus a

<small>Of power.</small>

normal and legitimate principle. It only needs to be regulated both in respect of degree and of its objects.

Emulation is a modification of the desire of power. It is the desire of power and of its increase relatively to others. It is a natural modification of ambition arising in social conditions. Right and proper in itself, therefore, it only requires to be moderated and guided both as to degree and object.

The desire of power carries with itself the desire of the means or the conditions of effectually exerting power—the desire of *possessions*, of *wealth*. This desire, when inordinate or misdirected, becomes *covetousness*, when it seeks, inordinately, wealth for use as means or condition—seeks in excess more money in order either to have more or to spend more freely—or *avarice*, when the inordinate desire is only to acquire and hoard, so that, if a ruling passion, it makes a man a miser.

Of knowledge. The desire of *knowledge* is another desire, seeking the perfection of the soul in respect to its intelligent nature. This desire becomes idle curiosity when it seeks not permanent growth of the mind, but rests satisfied only with the momentary gratification of the instinct for improvement belonging to the mind as a living nature. In this case, as is true of the other desires, when it disregards the end for which the desire was implanted, and cares only for the pleasure which is given by the Creator only to stimulate the propensity, it is diverted from its true design and becomes a wrong and an evil.

§ 95. Besides this large class of rational desires which are properly individual and personal, there is also the other class of rational desires, the social desires.

Social desires.

Highest among these is the generic *desire of society*, embracing divers modifications, such as the desire of kindred, of friends, of neighbors; the desire of social life in the family, the church, the State, in associations generally.

They are but the social affections, taking to themselves the peculiar characteristic of desires, grasping after their objects to appropriate them.

Subordinate modifications of the social desires, are the desires of approbation, of esteem, of applause, and the like, varying in respect to degree or to the mode by which others may be brought into agreeable or profitable relations to ourselves.

§ 96. Once more, the several desires may be combined with the expectation that they shall attain their objects. Hence we have those most important classes of desires, *hopes* and *fears*.

Hopes and fears.

Hope ever seeks a good, or a means of good, with expectation and desire of attaining it; while fear respects an evil or the occasion of evil, with expectation of its approach, but with desire to escape it.

Their objects are as diverse and as numerous as are the kinds of good to man that are possible or imaginable. They vary also indefinitely in respect of degree.

CHAPTER VII.

THE SENTIMENTS.

Sentiments.

§ 97. THE SENTIMENTS are feelings which are characterized either by intelligence or endeavor.

Described.

The mind is single with a plurality of functions. It is the same mind that feels, that thinks, that wills; and in putting forth either one of these functions, never entirely ceases from the others. Consequently, every mental state has somewhat of feeling, somewhat of intelligence, somewhat of volition or endeavor. It is so with the body. The body puts forth simultaneously the various functions of animal life. It breathes, it circulates its blood, it digests, it secretes, it receives and transmits sensations. No one of these vital functions need be suspended while the others are in exercise. But the mind is not to be conceived of as having its functions localized and separate in any such way or degree as the body. All that can be properly meant when the states of the mind are distributed into the several departments of feeling, intelligence, and will, is that these states are more prominently characterized at one time by one of these functions, at another, by another. When we say that a man, at a certain time, was in a passion, all we mean, is, that at that time, feeling predominated over intelligence and will, so that these latter functions, although by no means actively suspended, were overpowered and overshadowed by feeling.

These several states accordingly shade into each other, with indefinitely varying degrees of predominance of one function or another. It is, consequently, often difficult to say which predominates and gives character to the mental state.

Language hence does not accurately discriminate. We use the same term sometimes to denote a feeling, sometimes an act of intelligence, sometimes an act of will. The word *taste* thus is properly used to denote the simple feeling of beauty. It is also as correctly used to denote a judgment, a quick, accurate discernment in matters of art. It is still further used to denote an active governing disposition and pursuit. The word *love*, too, denotes a mere affection and also a determination of the will.

Two Classes. § 98. The sentiments divide themselves into the two general classes: 1. The *contemplative sentiments* which unite feeling with intelligence : 2. The *practical sentiments* which unite feeling and will, as the joint characteristics of the mental state.

The Contemplative Sentiments. § 99. The contemplative sentiments are recognized in language but to a very limited degree. The following will illustrate the general nature of this class of feelings.

Wonder is a sentiment which differs from the feeling of *surprise* in this, that it directly suggests an activity of the intelligence.

Esteem is a sentiment that with an affection for its basis, yet draws into it so as to give it character, a judgment.

Vanity denotes a sentiment with the feeling more prominent: *self-conceit* makes the judgment relatively more prominent.

Taste, as the term is frequently used, denotes a sentiment uniting a sensibility to beauty with accurate discernment of the elements of beauty.

§ 100. The practical sentiments, which may also be denominated moral sentiments, are more numerously recognized in language. Exemplifications are to be found in all the departments of feeling and voluntary action. They are found both as simple transient states of mind, and also as lasting dispositions.

<small>The Practical Sentiments.</small>

They spring up into exercise as well from impression on the sensibility as from determination of the will. The disposition of *generosity*, thus, may be called forth by an object of suffering need to which the purpose of the will at once responds and the whole soul is characterized by a feeling, willing act; or the settled purpose to do good may call forth the kindly feeling with which the act of kindness and good will is invested and made more acceptable and beneficent.

The two elements of feeling and willing combine, so to speak, in all proportions;—feeling sometimes greatly predominant, and sometimes the determining and purposing energy, and sometimes again, the two being in equiponderance.

The words used in language to denote these sentiments point to them sometimes as mere feelings, sometimes as acts of will. The term *pity* thus may be used to denote mere feeling—compassion—or it may be used to denote an act, ministering relief to want.

The explanation of all this is, that feeling and willing are functions of one and the same mind, and in these cases the two have such a natural connection and intimacy that each draws on the other. The will awakens and sustains the feeling; the feeling prompts and sustains the will. They act and react upon each other.

It may be added here that the intelligence just as freely enters into these states so that, were it of any practical utility, we might have a third class of sentiments in which, while feeling predominates and gives character to the mental state, the intelligence and the will are as truly concerned in it. The sentiments of *trust* and *faith* are thus seated in the intelligence—in the apprehension and acceptance of the true—in belief. They often at least imply a conforming will; and so while feeling may predominate and give character to them so that we rightly call them sentiments, the whole soul as feeling, knowing, willing, is concerned in them.

§ 101. The practical sentiments are most conveniently distributed into classes in reference to their respective objects, as the *Personal* sentiments which respect one's self; the *Social*, which respect our fellow men; the *Patriotic*, which respect our country; and the *Religious*, which respect God.

Four classes of Practical Sentiments.

Personal. Of the personal sentiments may be enumerated those of *modesty, humility,* and their opposites, *pride, arrogance; temperance, abstinence, asceticism; caution, presumption; candor, prejudice; zeal, heroism.*

Social. Of the Social Sentiments may be mentioned *probity, integrity, uprightness, fairness, equity, justice, fidelity; courtesy, urbanity, politeness; bounty, philanthropy, humanity, generosity, goodness; clemency, forgiveness, leniency, mercy, pity, compassion, condolence; homage, respect, disdain, scorn.*

Patriotic. The generic Patriotic Sentiment is *patriotism.*

Religious. Of the Religious Sentiments are *reverence, veneration, piety, godliness.*

CHAPTER VIII

THE PASSIONS.

The passions. § 102. THE PASSIONS are feelings in high degrees of excitement.

They are either transient excitements or permanent dispositions.

They are conceived of under divers forms of imagery. Thus under the imagery of heat they are denominated *glow, flame, ardor, fervor, fever ;* under that of force, *vehemence, agitation, perturbation, turbulence.*

They belong to all the classes of feelings which have been described. Thus, of Pleasure and Pain there are the passions of *rapture, ecstasy, transport, anguish, woe ;* of sensations, those of *voluptuousness, gluttony, sottishness ;* of emotions, *mirth, astonishment ;* of affections, *rancor, ferocity, truculence* ; of desires, *hankering, greediness, terror, horror ;* of sentiments, *squeamishness, dilettantism, vandalism, scrupulosity, punctiliousness, prudery, libertinism, abjectness, effrontery, brutishness, fiendishness, pietism, sanctimoniousness, sacrilegiousness.*

CHAPTER IX.

THE IMAGINATION.

The Imagination. § 103. THE IMAGINATION is the faculty of form.

Faculty of Form. We have recognized the sensibility in the larger sense, § 48, as the mind's function of receiving and communicating ideas, but in a narrower sense as limited to one part of this two-fold function—that of receiving idea. The other part of this function—that of communicating idea—is called the imagination. As the sensibility in this narrower sense is the capacity of Form, imagination is the faculty of Form.

Synonyms. § 104. This mental function is known under several different names, which designate different forms of its activity. The Greek language has furnished the name *phantasy*, contracted into the more familiar English word, *fancy*. This latter term use has made to point to more novel, or more playful exercises of the imagination. The Latin language has furnished the word *imagination*, etymologically signifying the imaging power or act. Inasmuch as the imagination constructs all its forms out of materials presented to it, the term *representation* has been extensively used to denote this power of the mind to represent to other minds, or to one's own, what has been presented to it. It is also called the *idealizing* power or function of the mind, and its products are accordingly called *ideals*. In this case, as also in the others, the name has been more or less restricted in use to some specific exercise of the general function. It should be added also, in respect to all these terms, that the use is loose and vague, and consequently more or less discordant.

§ 105. Of its proper nature and province it will not be difficult to form a satisfactory notion. If an orange be brought to the sensibility through the nerves of sensation—suppose those of sight—the sensibility as passive is simply impressed; the mind has a feeling of the object, which may be characterized as attended with pleasure. But the sensibility belongs to an active nature, and the mind in its active response to the impression, forms an idea of the object. The orange is now held by the mind in the form of an idea of its own, for such purposes and uses as it may determine—for contemplation, or for thinking, or for choosing and taking. It is, as an idea of the mind's own, in the mind no longer a mere impression, but an impression taken up by the mind's active nature and passed over and changed into an idea. It can be contemplated in

Explained.

respect to its form, as beautiful in its roundness, in its bright and mellow hue. It may be studied in thought and recognized as an object with certain attributes. It may be chosen and the choice go forth in endeavor to grasp and appropriate it. It may be set forth to others in language. The whole attitude of the mind towards it has changed from the mere sensuous passive impression. Then it received; now it puts forth. Then something came to it; now something is ready to go forth from it.

Not necessarily is the idea imagined identical in all its features with the idea received. The interpretation by the mind of the impression on the nervous organism may not be exactly conformed to the impression. Much less is the idea imagined necessarily identical with the object, the orange itself. The senses may have reported incorrectly through disease, through mingling with impressions from other objects, through the mind's own influence on the nervous organism. Still further, the mind shapes its idea of the orange not solely under the influence and control of the impression on the sensibility, but under the conjoined influence of all its own divers affections and dispositions at the time. Its idea of the orange may thus in many respects differ from the primitive object. The mind takes up the impression on the sense into its own activity, and then shapes it by its own states at the time — images or idealizes the object in them, through them, and by them. Often it becomes necessary for us to retrace our first impressions in order to correct or perfect our idea of the outward object.

<small>Ideals.</small> The Creator's idea of the orange, that is, the true idea of the orange, is accordingly in our minds more or less modified. It is somewhat differently shaped or formed. The image of it we have made varies in some respect from the original, and, in fact, from the impression on our nervous organization, for we have shaped it under the controlling influence of our existing mental disposition. The same orange is differently imaged in our minds according as we are occupied with other things or are entirely free to form a right and complete idea of it; according as we are sick or in pain, or well and in sound health; according as we are already prejudiced or are candid and unbiased. In short, the idea of the orange we shape into new form. The idea becomes by our mental action an *ideal*—something like the idea, yet somewhat different.

If we now examine this ideal we at once discover three elements that have entered into it. In the first place, there is the idea itself more or less perfectly impressed on our nervous organism and taken up through this impression more or less perfectly into our own minds. In the next place, there is our own mental state in which this idea is formed; in which it is incorporated, and which has thus become a kind of body to it. In the third place, there is our own mental activity which has taken up the idea and incorporated it in this body

Such is the true nature of an ideal, as carefully

distinguished from an idea, in its most abstract and shadowy form, and in the last analysis. This is the first work of the imagination; it forms, it creates an ideal. This ideal may be of any outward sensible object, as an orange, or of any spiritual object; of any material or mental thing, and also of any attribute of such a thing or of any relation of it; of whatever, in short, may be apprehended or experienced by the mind.

This primitive ideal, the first work of the imaginative faculty, is, however, susceptible of being still further shaped and formed. The imagination by the instincts of its own active nature is prompted to proceed to shape this ideal in its sensuous organism, or in its own interior thought, or in both. We have accordingly two kinds of these secondary ideals or images or forms. We have thus the following definitions and classifications:

§ 106. An ideal is a product of the imagination. It consists of three elements, —idea imagined, matter or body in which it is imaged, and the act of the imagination itself embodying the idea in the matter.

Defined.

§ 107. Ideals are primitive or secondary. A PRIMITIVE IDEAL is the first product of the imagination, embodying the idea in the mind's own furniture of thought or feeling or purpose.

Primitive ideals.

A SECONDARY IDEAL is ideal shaped in some new matter or body.

Secondary ideal.

§ 108. Secondary ideals are of two kinds: 1. Those which are shaped in

Sense ideals.

the sensuous organism—*sense—ideals*

2. Those which are shaped in the pro-
Spiritual ideals. per spiritual or mental matter or body—
spiritual ideals.

It should never be forgotten in the study of mental phenomena that these phenomena belong to the same one, indivisible mind, which never perhaps surrenders its whole activity to a single one of its divers functions, but whose acts and states are ever more or less complicated. When we therefore speak of the mind as having a sensation, as of sight, of cold, of fatigue, we do not mean that it may have not other sensations, other feelings, perceptions, judgments, purposes and endeavors at the same time. We only mean that the mental state is characterized at that time by the sensation as an actual, perhaps, predominant element of its state. Thus simultaneously with the sensation, with the sight of the orange, the mind may desire it, may perceive it, may recognize it as round and as sweet, may purpose to take it, may put forth endeavors for it. But in order to mark all the divers features of this complex mental phenomenon of the imagination, and to understand it as one whole, we must note separately (1) the sensation or the impression on the bodily organism ; (2) the primary ideal which the mind by its own native activity forms from that impression ; and (3) the secondary ideal which it forms in its own furniture of feeling, thinking, willing, which secondary ideal it perceives and judges and chooses as the mental image of the object, and also impresses on the brain with its nervous apparatus in its endeavor to take the object. All these elements enter into the phenomenon and constitute it what it is. They are essential to it, so that we cannot conceive its possibility if any one element were taken from it.

Of the primary ideal, it may be said here, that while it is to be discriminated, on the one hand, both from the affection of the sensuous organism, as widely as a mental act from a bodily state, and also from the idea conveyed in that affection, and, on the other hand, from the mental activities that are grounded upon it, such as thought and purpose, and must be enumerated in any complete analysis of mental phenomena, it does not seem to require of us any further statements in respect to its character and relations than those already given. As to its elements, it possesses the three that enter into every product of the imagination—idea, matter, imagining act. As to its relations, it is the primary mental act from which all other mental activities spring and on which they all rest. It is variously modified of course by the character of the impression which originates it; by the character of the mental furniture in which it is formed; and moreover by the vigor of the imagining act itself.

CHAPTER X.

THE IMAGINATION—SENSE-IDEALS.

Definition. § 109. SENSE–IDEALS are products of the imagination shaped in the bodily organism.

We have recognized a twofold structure in the nervous organism—one for receiving, the other for putting forth ideas; a twofold system of nerves,—one *afferent*, otherwise called *sensory nerves*, bringing to the mind; the other *efferent*, otherwise called *motor nerves*, carrying from the mind.

The imagination uses this second system—the efferent or motor system—for its instrumentality, as the sensibility in the narrower sense uses the first,—the afferent or sensory system.

The mode of connection between the mind and the material system of the brain and nerves is wholly wrapped in mystery; and we are utterly unable to explain either how the brain carries ideas to the mind, or how the mind conveys outward its ideas through the brain. The keenest anatomy cannot discern the point of this connection. No science, indeed, can tell whether the connection is at a single point or over an extended portion of the organism.

It is accordingly entirely inexplicable how it is that one state of the mind, one ideal, should be followed by a motion of the hand, and another by that of the tongue or lips. All that we know is the simple fact, that we think, we imagine, we form ideals, and instantly the nervous organism repeats the act, and this or that nerve, this or that nervous centre, this or that part of the brain, responds. We know that excessive mental action, particularly excessive exertion of the imagination, by putting the brain, or some part or other of it, into movement, induces weariness, and, perhaps, disease, and ultimately death; and that injury to the body—to the nerves, to the brain—reacts upon the mind and disturbs or even diseases its action. This mental action, at first affecting the brain proper, may ultimately reach even the portion of the nervous system not directly connected with the brain, the ganglions or nervous centres from which nerves go out into the respiratory, the circulatory, the digestive systems of organs. A mere recollection, for instance, of some tragic scene, of some danger encountered, of some wrong done, sometimes suspends the breath, quickens the pulse, moves a sigh or a sob, disturbs all the alimentary functions.

§ 110. The affections of the sensuous organism by the sense-ideals vary indefinitely in kind or character, and also in degree.

Modifications.

They vary with the kind or character of the ideal itself. An ideal of a visible object, as of an orange, affects the brain and its nervous retinue differently from an idea of a sound; an ideal of a picture differently from that of an action. Different parts of the brain, different nerves are brought into exercise in the different cases.

They vary with the condition of the body and particularly of the nervous organism. A diseased body may make the imagination even of a generally agreeable object disagreeable or offensive. The imagination of dainty food nauseates in sea-sickness. We are credibly informed that a man who had been wrapped when sick at sea in a cloak, could not wear it afterwards on land without the return of the nausea with which it had been associated.

They vary in degree with the energy of the imagining act, and also with the susceptibility of the organism. A vivid imagination may quicken the blood, suffuse the cheek, brighten the eye, fill with animation the whole frame; while a dull imagination, even when framing an ideal of the same object, may not sensibly stir a fibre of the body. A susceptible organism, too, moves quick and strong from an ideal that would scarcely stir a dull and gross sense.

Physiologists generally agree in the opinion that all mental action is attended by some change in some part of the brain-system—the *cerebrum*, the assigned seat of higher mental or proper rational action; the *cerebellum*, the supposed seat of muscular exertion; or the *sensorium*, the seat of the special senses, which has only afferent and no efferent nerves. Three different opinions are held in regard to this relationship of mind and body. One is, that all kinds of so-called mental action are caused by the organic changes in the brain or nerve system—a pure materialistic dogma. A second is, that the brain-change is a necessary condition of any mental act, but that the mind is a distinct, separate agent from the bodily organism, the chief, if not only characteristic of which is will or power of volition or self-determination. A third is, that the mind is, after at least the first awakening of its activity by a sensible change, a self-active power, capable of originating action while yet its action is ever attended by a change in the brain. This opinion, it would seem, is the only one compatible with the recognition that mind and matter are distinct entities, and that mind is essentially active while matter is essentially passive. In the mysterious union of mind and matter in man we find that a change in either of the constituents is ever attended by a corresponding change in the other.

§ 111. Sense-ideals may be accompanied or not by conscious intelligence and exertions of the will.

Relations to intelligence and will.

There may be no reason to doubt that every mental state, every act of the imagination, every ideal has its proper influence on the brain. But it is certain that often the intelligence fails to take any account of these impressions. All unconsciously, the feet, the fingers, the eyeballs, move in mere contemplation, in simply beholding a beautiful picture or in hearing musical sounds. Then, on the other hand, we often take up into thought these ideals and study their character, their attributes, judge of their accuracy, or their completeness, or their fitness

In the same way these sense-ideals often move our bodily frames with no control, no help, no hindrance from the will. Then again our wills embrace these ideals, and we infuse into them a new energy; we guide them, or we repress and hinder them. An idea of commendation or of deserved reproach often suffuses the cheek with crimson against all power of the will to hinder such effect; and often the blush appears as unconsciously as involuntarily.

These general statements of the reciprocal influence of the mind and the bodily organism we will illustrate by some well attested facts. They will be presented under the several classes of Phantoms; cases of Exalted Sensibility; and instances of Suspended Sensibility.

§ 112. I. A PHANTOM is a sensation, *Phantoms defined.* produced not by an external object, but by an impression from the mind—from the imagination—or the sensuous organism.

Here there is a real affection of the organism, but the cause is not from the world exterior to the body but from the mind. The impression on the organism is reported back to the mind just as if the impression were from without; and, therefore, it appears to the mind precisely as if an external object had made the impression.

Sir David Brewster, in his letters on Natural Magic, narrates the case of a lady of high character and intelligence, whose vivid imagination so affected her nervous organism as to occasion frequent and very striking illusions. She heard unreal voices, as that of her husband calling her by name to come to him, repeatedly, distinctly, and loudly. One afternoon, on entering the drawing-room, she saw, as she supposed, her husband standing before the fire and looking fixedly at her. Supposing he was absorbed in thought, she sat down within two feet of the figure. After two or three minutes she asked him why he did not speak to her. The form then moved off towards the window at the further end of the room and disappeared. The appearance was in bright daylight, and lasted four or five minutes. At another time, sitting with her husband in the drawing room, she called his attention to what she supposed to be the cat. She pointed out to him the place where the phantom was; called the cat to her; when trying to touch it she followed it as it seemed to move away from her. At another time she saw a favorite dog apparently moving about the room, while she was holding the real dog in her lap.

A similar case, equally well authenticated, is that of a lady who, while seated by a table, saw the figure of a man enter the door opposite, and move slowly towards her, and then distinctly heard him say that he was come from the Spirit-world, charged with a message to her, which he then communicated, solemnly enjoining it upon her to do what was required. The form passed slowly by her around the table and vanished by the window on the opposite side of the room.

In these two cases, there had been disease which had affected the nervous sensibility. In each case the senses of sight and of hearing were both concerned.

Such spectral illusions are, in fact, not infrequent in fevers. The writer, in the approach of a febrile attack, at intervals when free from delirium, imagined the phials of the medicine closet in the room to be men and women of the most grotesque and fantastic shapes and movements. They seemed as real as the doorway and the shelves on which the phials stood. His nervous system, in some part, was affected just as such real objects would have affected it in order so to impress the mind.

The cases already instanced were cases of involuntary imagination. The late President Hitchcock of Amherst College, relates his experience of similar illusions which, in part, and particularly at first, took place without any design or expectation of his, but in part, and subsequently, were occasioned and induced of express purpose. He was able, by bandaging his eyes and thus entirely excluding the light, to bring before his mental vision images of various kinds of objects and scenes as distinctly and as vividly as if realities. Having thus covered his eyes on one occasion for the purpose of experiencing these visions, he reported what passed before his view successively to one who took down the reports thus : "The space around me is filled with huge rocks moving past me in all modes, full of caverns, but too dark to be well seen ; they hang over me now and look splendidly ; some of them appear to be serpentine. Some of these rocks seem a hundred feet long. Against the side of a wall I see three young ladies sitting and laughing ; lighted candles are before them, and chains, machinery, etc., around them. I lie in a vast cavern; the rocks are rolling around me like clouds; they are within a foot of my face; some are sandstone and some granite. I have a glimpse into a large city; but a carriage-maker's yard, full of rubbish, almost entirely obstructs my view." This is but a brief extract from his account of these phenomena which occurred during an attack of fever, in which, however, there was no tendency to mental derangement.

§ 113. 2. EXALTED SENSIBILITY. The sensibility sometimes exhibits extraordinary tenderness and life. This occurs most strikingly when both the bodily organism is unusually excitable and the imagination is also unusually vigorous and active. Well authenticated facts explain to us much that might otherwise seem to be the effect of supernatural agency.

Exalted Sensibility.

The case of Jane C. Rider, of Springfield, Mass., related by her physicians, is one of many, but one of remarkable interest. At intervals during several months, in a great variety of circumstances, she could, at night, or in a darkened room, and with her eyes closely bandaged, distinguish by her eye all ordinary objects presented to her. She at one time read, with her eyes thus bandaged, audibly and correctly, with some hesitation however at the most difficult words, nearly a whole page from a small volume handed her. The distinguished physicians, who observed and narrated the case, correctly ascribe the result to "the combined effect of two causes; first, increased sensibility of the retina, in con-

Exemplified.

sequence of which objects were rendered visible in comparative darkness ; and, secondly, a high degree of excitement in the brain itself, enabling the mind to perceive even a confused image of the object." We must interpret " the excitement in the brain enabling the mind to perceive a confused image," here spoken of, as not in the body, but in the mind itself, whose imaginative function was in a state of exalted vigor.

The most frequent exemplifications of this state of exalted sensibility occur in cases of fever and of *delirium tremens*. The power of the imagination over the nerves in this last named disease is almost incredibly great. Robust, stout-hearted men, even men who had seemed hardened and callous to every impression, reckless and fearless of every thing, in this disease see visions and hear sounds that only the pit of despair can know as realities, and strong frames sink down in the course of a few years to death under the horrors of an excited and uncontrolled imagination.

§ 114. 3. SUSPENDED SENSIBILITY.

Suspended sensibility. The more normal and familiar phenomena of this class occur in ordinary sleep. The characteristic feature of sleep is

Sleep. the partial suspension of the reciprocal action of the mind and the body on each other. This suspension, in healthy sleep at least, is never entire. As sleep comes on, one sense after another in quick succession becomes inactive. The order varies ; but the hearing and the touch are generally the last to sink into repose

Commonly the nerves of sensation and the nerves of motion cease their functions almost simultaneously. The eyelids droop, the head sinks, the limbs drop to some external support, while, nearly at the same time, the taste, the smell, and the sight first, and then the hearing and the touch, suspend all communication between the soul and external things. All the vital functions, nevertheless, as those of respiration, circulation, nutrition, secretion, and absorption, go on as in wakefulness. The heart, however, beats slower, and the breath is less rapid, and in early life absorption and nutrition are more active. The brain collapses from the diminished flow of blood into it.

Sleep is more or less profound, the suspension of the connection between mind and body is more or less complete in different persons and also in different conditions, internal or external, of the same person.

Facts abundantly show that one sense may be fully awake while others are asleep. A nurse, watching the sick, will wake on hearing the striking of the clock, or on hearing the slightest call of the patient. Erasmus relates of his friend Oporinus, a celebrated professor and printer of Basle, that after a wearisome journey with a bookseller, he undertook in the evening at the inn to read aloud a manuscript about which they had been conversing during the journey. The bookseller discovered after a time that Oporinus was asleep while he was reading. A like experience befell the writer who, after an exhausting journey by night and day, undertook to read to others a long document of much value and interest with which he had become familiar during his journey. He fell asleep, but continued reading till, after a page or two, the hand which held the manuscript dropped and awakened him. The sight in these cases remained awake, as also the motor-nerves concerned in reading, while other senses were asleep. Sir William Hamilton relates the case of a postman who daily traversed, on foot, the route between Halle and a town some eight miles distant. Over a part of the route which lay through a meadow, he generally slept; but on coming to a narrow foot-bridge, which was to be reached by some broken steps, he uniformly awoke. Soldiers, it has been often observed, wearied by a long march, sleep while their feet move on as when they were awake.

§ 115. DREAMING is a familiar phenomenon of sleep. Ordinarily we include in the notion of a dream that of a connection between mind and body, reciprocally acting upon each other. But a right explanation of this interesting phenomenon involves the truth of the continued activity of the mind even in what we call profound and perfect sleep. The mind, as we have seen, is essentially active. To cease its activity, for it, is to die, since action is its very life. The life of the body even ceases when all action in it ceases, when circulation and respiration and secretion and absorption cease. Certain modes of thought or feeling may

Dreaming.

Mind ever active.

be suspended ; but to conceive of all thought and feeling and willing as stopping, is to conceive of an extinct soul. There is no evidence that the mind wholly suspends its action in the profoundest sleep. That we cannot recall the thoughts we may have had in sleep does not prove that we did not think. Let one give himself to musing for a half-day, letting his mind rove uncontrolled in any direction and towards any object that may offer ; he will, in all probability, be unable at the close to recall one in a hundred of the objects that have flitted before his mind. The mind is active when it loses itself, as we say, in sleep—when it falls asleep ; it is active when it recovers itself to wakefulness ; it certainly is sometimes active during sleep, as what we can recall of our dreams evinces ; who can suppose it ceases action in sound, undreaming sleep, more than in those wakeful hours, the flying thoughts of which wholly escape our recollection? We say loosely we are not conscious of thinking or feeling during our sleep. If we mean that the mind was not conscious when acting, this is to mistake utterly the essential attribute of mind which is necessarily conscious of all its own action. If we mean that we are not now conscious that we had any feeling or thought while we slept, then we mean only that we are unable now to recollect--to bring into our present consciousness the fact that we thus thought or felt. Still further, there are curious facts which make this supposition, that the mind may be active, and therefore consciously active, even during the profoundest

THE IMAGINATION—SENSE—IDEALS.

sleep, extremely probable. There are many well accredited facts showing that the mind not only acts in sleep in ways that of itself it **Proved by facts.** was utterly unable to recall, but also sometimes acts with an energy and intensity beyond what it ever knows in wakeful hours. A mathematician, who had long labored in vain to solve a mathematical problem, one morning found the solution on his table. He had risen in his sleep and worked out the solution, but of the operation he had no recollection, and the only evidence that could convince him of his dream-work was the paper on his table. Franklin was wont to find in the morning political questions that had tasked his wakeful hours the day before clearly resolved in his mind. Coleridge dreamed out his poem, "Kubla Khan," while asleep in his chair. He wrote out from recollection immediately on waking what appears of the poem in his works, but being interrupted lost the power to recall the rest, which he yet believed he had fully composed in his dream to the extent of three or four times what he had written.

Similar to this experience of Coleridge is that related in MacMillan's Magazine in 1870, of a lady who had been pondering during the day on the many duties which "bound her to life." She dreamed her feelings into verse, and on awaking was able to recall the following stanzas

> Then I cried with weary breath,
> Oh, be merciful, great Death!

Take me to thy kingdom deep,
Where grief is stilled in sleep,
Where the weary hearts find rest.

* * * * * * *

Oh, kind Death, it cannot be,
That there is no room for me
In all thy chambers vast;
See strong life has bound me fast,
Break the chains, and let me free.

But cold Death makes no reply,
Will not hear my bitter cry;
Cruel Life still holds me fast.
Yet true death must come at last,
Conquer Life and set me free.

Dr. Carpenter, in his mental physiology, relates an occurrence which proves not only that the mind may be capable of more intense activity in sleep than in wakefulness, but also that a protracted mental operation of the highest character may take place in sleep of which no adequate recollection survives on waking. A man was called to compose a discourse for public delivery on a set occasion. He gave himself to the effort, and the evening before the appointment was to be met, he had composed something, but lay down utterly disgusted with his performance. He fell asleep and dreamed of a novel method of handling his subject. When waking he rose to commit his new thoughts to paper, but found to his astonishment on opening his desk, that they were already written out, the ink being hardly dry.

Of the greatly increased activity of mind sometimes experienced in sleep, we have, indeed, manifold illustrations. The following may be added to the instances already given : A person, aroused from sleep by some water sprinkled on his face, dreamed of the events of an entire life before coming to full wakefulness. There is an accredited record of an officer awakened by the morning gun, who dreamed of hearing an alarm-call to battle, of rising, equipping himself, going to the field, marshaling his men, engaging in a long and doubtful battle and of driving the enemy from the field, every step as orderly and as complete as if all real, yet dreaming through all this before the reverberations of the gun had died away on his ear. De Quincy says of his mental activity in his dreams that he sometimes seemed to have lived seventy or a hundred years in a single night.

Sympathy of body and mind in sleep. The mind thus never in sleep entirely dropping its activity, is more or less in sympathetic connection with the body. A patient in a hospital in France, who had lost a portion of the scalp and of the skull, thereby exposing the movements of the brain, was observed in calm sleep to exhibit a motionless brain, but in a sleep disturbed by dreams to be in proportionate agitation. It would be rash to infer from the apparently motionless brain in calm sleep that the mind itself was also inactive; but the agitation of the brain at times evinces the fact of the continued interaction of mind and body in sleep. This motion in the brain occasioned by the mental action, may take place interiorly so as not to show itself at all on the surface ; it may extend throughout the entire structure of the brain; it may extend farther, into the nerves that issue from the brain, the sensory and all the motor nerves; it may reach a part or the whole of the entire nervous organism. Dreams often occasion movements of hands and feet; sometimes of the organ of speech. A dream of fright will occasion sudden convulsive bodily movements, as if to avert or escape danger. Dreams often occasion sighs and groans and outcries of alarm, or smiles and audible laughter. Some persons talk frequently in their sleep. Conversation can sometimes be carried on to some considerable length with them. The writer knew a student in college who acquired the art of leading his roommate when asleep to translate his Greek lessons for him night after night. An English officer was led in his dreams by his companions, who were aware of his peculiarities, to go through the whole process of a duel, and was awakened only by the report of the pistol which he fired in the supposed combat.

The bodily organism acts upon the mind during sleep, as does the mind upon the body, in modes and degrees variously modified. A bright light brought into the room where one is sleeping, or a noise or a touch, there is reason to believe, often influences the mind and shapes the dream. Dr. Gregory having placed a bottle of hot water at his feet dreamed of going to Mount Etna and of extreme heat. In the same way the disturbance of the vital functions, or any pain in the body, often occasions distressful dreams. A posture of constraint in which the mind becomes conscious of inability to command the muscles, gives rise to incubus or night-mare. The mind, conscious of this inability to move for defense or for escape from the danger which the constrained posture of the body had occasioned, suffers the extreme anguish and horror of one in real danger from which he sees no way of extricating himself. He is in the mental condition of one whose limbs are inextricably entangled in the burning wreck of a railway train, and who sees the flames steadily and irresistibly moving upon him.

§ 116. Besides the normal phenomena of suspended sensibility in sleep, there are the abnormal states of *catalepsy* and *somnambulism*.

Catalepsy. In CATALEPSY, the subject seems like one in quiet sleep, with regular pulse and respiration, but beyond the reach of all the ordinary excitants from sleep. Intense flashings of light on the eyes, loud noises, pungent odors, punctures of the skin, shakings of the body, severe blows and bruises, prove of no avail to restore to wakefulness. A variety of this affection is *ecstasy*, occasioned generally by religious excitement. In these cases the mind continues active, although the connection with the senses is more or less suspended. Sometimes only a part of the nerves seem to lose their functions, as is shown in the case of persons who have been supposed to have died under these paroxysms and have been laid out for burial, who yet continued conscious of all that passed and on recovery repeated what was said by the attendants. More frequently, however, the memory fails to recall what has passed during the attack.

Yet radical and permanent changes of disposition and character, as from dissoluteness and irreligion to soberness and piety, which are known to have attended these experiences, show that, although impossible to be recalled, there must have been clear, strong thoughts, deep feelings, decided purposes.

§ 117. Somnambulism. Somnambulism is a form of partially suspended sensibility, combined with more or less exalted susceptibility in some of the senses, and particularly with a controlling activity of the intellect and will, reaching to the bodily functions. This is indeed the special characteristic feature of somnambulism, distinguishing it from dreaming and from catalepsy; the somnambulist prominently manifests a use of the bodily organs for some set purpose or object. In dreaming, the control of will is relatively dormant; the mind is floated along hither and thither without guidance of its own. In catalepsy the mind may exercise its reason and its will, forming purposes that shall be permanent and govern the future life, but its action does not go out into the bodily movements. In somnambulism this last is the characteristic feature. The somnambulist is "a dreamer who is able to act his dreams." He rises from his bed and walks the street, or climbs to the top of the house, passes quickly along dangerous ways, or delivers an harangue, or recites poetry, or works out mathematical problems, or executes works of art. The affection proceeds from a highly excitable nervous organization, which may be stimulated either by some mental act or by some affection of the bodily system in ordinary health or in disease, or even by artificial appliances.

Many instances of this phenomenon are on record. They have been noted from the earliest times; they were described by the ancient Greeks. As we should suppose beforehand, they are diversely modified. The mental activity is sometimes most marvellously stimulated so as to transcend all ordinary experience; sometimes it is only of the most usual degree; sometimes one function of the sensibility is suspended while another is exalted to an extraordinary degree, or, it may be, remains only of the usual energy; sometimes the experience during the somnambulistic attack is remembered as in dreams; sometimes it is beyond recollection while in the normal condition, but it may be revived again fresh and vivid when the attack recurs, so that the subject seems to live two lives, remembering in the normal state only what has occurred in that state, and in the somnambulistic state only what has passed in that.

Several cases will be narrated to illustrate these general characteristics. The Archbishop of Bourdeaux relates that a young ecclesiastic was in the habit of getting up night after night, and, while giving conclusive evidences of being asleep, going to his room, taking pen, ink, and paper, and composing sermons. When the Archbishop placed a piece of pasteboard between his eyes and the paper, he wrote on, not seeming to be incommoded in the least.

Gassendi reports that a somnambulist used to

rise and dress himself in his sleep, go down to the cellar and draw wine from a cask, seeming to see in the dark as well as in full daylight. He answered questions that were put to him. In the morning he recollected nothing of what had passed.

Colquhoun relates that a young woman of twenty years of age, frequently passed from a state of proper catalepsy into that of somnambulism. She sat up on the bed and spoke with an unusual liveliness and cheerfulness, and in continuation of what she had spoken in her previous fit. She would then sing and laugh, spring out of bed, pass round the room, dexterously shunning anything in her way, then return to her bed and sink into the cataleptic state. All means tried to awaken her were ineffectual, such as burning a taper close before her eyes, pouring brandy and hartshorn into her eyes and mouth, blowing snuff into her nostrils, pricking her with needles, wrenching her fingers, touching the ball of her eye with a feather and even with the finger. When informed of what she had done, she manifested deep mortification, but never could recollect anything that had occurred.

Cloquet reported to the French Academy the case of a lady who, having been thrown into somnambulism by some artificial means, had an ulcerated cancer removed without manifesting the slightest sensibility. She was kept in the somnambulistic state for forty-eight hours, and so completely that when she awoke she had no idea of the operation till she was informed of it. She talked during the attack calmly and freely about the operation when it was proposed to her, notwithstanding she had shrunk with horror from it when awake, quietly prepared herself for it, conversed with the operator during the operation, without any motion of limb or feature, or any change of respiration or of pulse evincing that she was sensible of pain.

In Massachusetts, some years since, a girl of fourteen years of age, of a nervous temperament, but without any extraordinary intelligence, after having fallen asleep in the day time, would rise from her chair and deliver a sermon, which she preceded by the usual religious services, as if to a large audience. These discourses, which far transcended in mental power her wakeful ability, she would deliver day after day or on alternate days, without repetition, however, of thought or language.

Another interesting case of somnambulism is that of a young lady who became a competitor in a school in which prizes had been offered for the best paintings. As she returned in the morning to her work, she repeatedly observed that alterations beyond her own skill had been made to the painting. She charged her companions with the interference, and when they denied it she took precautions to prevent further interference with her work. Her own movements were now watched, and she was seen to rise in sound sleep, dress herself, go to her table and work on her painting. The prize was given to her, but she was loth to receive it, as she insisted that the work was not her own.

CHAPTER XI.

THE IMAGINATION—SPIRITUAL IDEALS.

Spiritual ideals. § 118. SPIRITUAL IDEALS are products of the imagination, shaped in the mind's own furniture.

This class of ideals may have been originally derived from sensation, more or less; they may come to be shaped in sensuous forms. But they may exist at the time independently of all sense connecnections. It may be that in producing them the working of the imagination may be connected with movements in the structure of the brain in its most interior part, or in any more external portion of the nervous organism. It is not improbable that any mental activity may, from its mysterious alliance with the body, draw in with it also some movements of the bodily organization. But it is clear that we may conceive of a purely mental act separated from all sensuous elements. Such a mental act as formed in the mind by the imagination is a spiritual ideal.

Source. § 119. These ideals are all formed out of the mind's own possessions—out of the stock of thoughts, feelings, and purposes which it has in itself.

They are not made from nothing. Their variety, richness, greatness, depend on the growth and the attainments of the individual mind. A child's ideas are simple, narrow, meager, compared with those of a mature, cultivated mind.

Of this stock of material out of which the imagination forms its ideals it will be important to obtain a fuller and clearer understanding. If one were to be asked in regard to a journey he had made during a preceding year, he would be able to answer so as to convey some idea of it ; as, we will assume, in what month he set out; how long he was gone ; what places he visited ; what objects and scenes most interested him. All these ideas of his journey which he thus communicates in his answer are the products of his imagination, which, entering into the stock of his recollections, shapes its ideals out of them. These ideals thus formed go out, as he communicates them, through the sensual organism in sounds, in words, which the inquirer on receiving them garners into his stock of ideas or mental possessions. This complex act of taking out of the stores of the mind's ideas such as would meet the demands of the inquiry and of shaping them in ideals to be then expressed in words. Sir William Hamilton, with a nice analysis, has explained as involving the exertion of a threefold faculty, (1) the *memory* proper, the retentive or conservative power by which the mind retains its ideas ; (2) the *reproductive* power by which the mind calls forth what was lying dormant in memory ; and (3) the *representative* power by which the mind holds up before itself the ideas which it has reproduced from memory. Whatever may be thought of the propriety of recognizing these distinct faculties—retention, reproduction, and representation—as faculties of the intelligence, it is clear that we have this three-fold phenomenon to recognize and explain; first, we have the fact that the mind retains its ideas; secondly, that out of such retained ideas it frequently calls forth this or that for its use; and thirdly, that it shapes such recalled ideas into new forms for communication to others or for its own study. It is obvious, moreover, that the retention and the reproduction into present consciousness of ideas are the two necessary conditions of representing or imagining. We shall therefore in order consider these two conditions of ideals—memory and reproduction—in separate chapters, reserving for a distinct chapter some additional explanation of the imagination itself as an idealizing power.

<small>Bodied in ideas.</small> It remains to be observed that these spiritual ideals are not only shaped out of the mind's own stock of ideas, but are also shaped in them.

The recollections of a journey shape themselves very differently at different times. If one has observed the Parthenon of Athens, and should in after years recall and represent his idea of it retained from the impressions made upon his mind at the time of observing, his account of it would be different in some particulars, if given the first year after his return, from that which he would give the tenth. Some details would in this latter instance have slipped out of his ideal; the others would be more or less differently arranged, and the several features would stand out in different degrees of prominence relatively to the others. His account, and consequently his ideal, moreover, would vary with the design or end for which he recalled it. To describe the Parthenon to a child, he would shape his ideal in one way; to a cultivated artist, his description would set forth his ideal shaped in quite another way. But in every case his imagination shapes its product in the mental furniture of the time. It is outlined in existing feelings, thoughts, and intentions. It is not only outlined in them and bounded out in and by them, it is also colored by them. His ideal will be at one time glowing with the feeling which transports him at the time of describing; at another, it will be dull and dim, as his mind at the time is heavy and clouded. The same idea of the Parthenon thus will be embodied in the varying experiences of the hour and assume a form corresponding to them. The character of the ideal, the distinctness of its outline, the perfectness, the completeness and the richness of the rendering, will also vary with the vigor of the imagination at the time and with the design for which it acts.

CHAPTER XII.

MEMORY.

§ 120. By MEMORY, in its stricter sense, is meant simply the retentive attribute of mind. The best view to take of memory is to regard it as the holding on of a feeling, a thought, or a purpose in the continuous life of the soul, § 24. Every impression made upon it abides in its effect; every thinking act continues, never becoming extinct; every choice and purpose likewise remains a part of the mind's ceaseless activity. We may as well suppose that matter or force can be annihilated as that the effect of force can die out utterly; and so we may as well suppose that the mind or a part of it may die out, as that its action, any movement it may experience either from the impressions of other forces or from its own prompting, may utterly cease to be. We easily enough accept the truth that strong feelings, momentous thoughts, decisive purposes of our lives, may live on for ever; we cannot with any consistency hesitate to believe that less important acts of our minds also live on. If a great thought has a life that

[margin: Memory. Explained.]

reaches through the entire life of the mind, every lesser thought must have the same perpetuity. The single drop, as well as the great tributary, remains ir the swelling river. The great tributary of thought is in fact made up of the little drops of experience, and cannot be without them.

The impossibility of recalling all the transient thoughts of past years, does not disprove this supposition of the continuance of every thought and feeling. This impossibility is to be attributed to the limited power of the human mind to recognize the minute parts of its experience, not to the annihilation of those experiences. The originally clear stream of the Mississippi receives into its volume the whitish muddy waters of the Missouri, then the greenish muddy Ohio, and then the reddish muddy streams of the Arkansas and Red rivers. For a little space each tributary maintains its separate integrity so far that it may be distinguished; but as the augmented stream rolls on, the waters intermix more and more, till in the lower course of the river the several discolorations seem to our limited vision to be all blended into one mass of turbid color. But each particle can by an infinite mind be traced back to its source, and the whole volume of water in the channel is what has come into it from these separate sources. In this case, indeed, some of the original supply is wasted into the air by evaporation, by diversions into little lakes, by use for irrigation or other purposes; but in the great current of the mind's activity, nothing can be supposed to be thus wasted. All that has entered the stream, the contribution of every minute transient experience, remains to swell and to characterize it. We have thus this great law of mind in relation to its power of retaining its function of memory.

<small>Perpetuity of mental activity.</small> § 121. EVERY FEELING, EVERY THOUGHT, EVERY CHOICE, ABIDES WITH THE MIND FOREVER.

The proofs of this principle of memory may be summarily exhibited as follows:

<small>Proof 1, from presumption.</small> 1. The presumption is that every action of the mind continues. It may not continue entirely unmodified; its form may change; it may exist as cause or in its effect; it may be now more or less connected with one mental experience and then with another; it may be variously colored or shaped thus in the progress of experience. But as we must believe that everything that is, continues, unless we have some reason for believing that it has ceased to be, and as there is no such reason for supposing our mental action to die out utterly, we must accept the law of the deathlessness of memory as valid.

<small>2. From analogy.</small> 2. Analogy confirms this view. Matter, we believe, is never annihilated; force is never annihilated; motion, the effect of force, is never annihilated; we conclude that, unless something can be shown to destroy the analogy, mind and its action continue. Matter changes its form; force changes its direction and also its form; one motion passes into other motion, as the motion of gravity or of the mass passes into the motion of cohesion and repulsion, the mo-

tion of atoms; but with change of form each continues. The quantity of matter in the universe, the quantity of force, the aggregate of the quantities of motion remain the same. At least created things have no power to destroy their own being or their own essential attributes. We are led thus to believe that mental activity once originated abides in some form, positive or negative, as long as the mind itself exists; that every feeling, thought, and purpose hold on and are retained in the mind's being.

3. From common experience. 3. Facts from ordinary experience strengthen these arguments from presumption and from analogy. It frequently happens that little circumstances, which we should have supposed were too trivial to be retained in memory, reappear in our thoughts, called forth by some association perhaps strange to us. Objects which we have seen, words which we have heard from others, or had uttered ourselves, that had all vanished from our consciousness, somehow come up into our thoughts afresh. In old age little circumstances that occurred in childhood are recalled with a freshness and a vividness that seem surprising. Sometimes all the great experiences of middle life have faded out from the memory of the old, while the scenes of childhood are revived, and are lived over in recollection with wonderful exactness and fulness. In the same way, too, that which we have dreamed and which had so lightly impressed us that we did not recall it when we waked, returns, months or years after it may be, in second dreams that recall even the little details of the first. Still further, we have the great fact that thoughts and feelings and dispositions are perpetually coursing through our minds, which could appear there only as the retained acts of previous life. These thoughts and feelings may not come up, and for the most part do not come up, into distinct consciousness one by one. But there is a volume of thought that is retained from the past, streaming along and shaping and coloring our present thought. We meet, for example, an old friend in the street after a long absence; thoughts, feelings, scenes, objects, pleasures, sorrows, plans, hopes, actions, that have lain buried for years out of conscious thought, pour through our minds. In truth, every thought we have must be affected more or less by every thought we have ever had, really, although it may be imperceptibly to our finite vision. The little boat that floats on the broad bosom of the great river near its mouth, is sustained in its part by every drop that has come into the stream from the most distant little spring from the other side of the continent. We are unable to discern any lifting of the water except for a few inches from the boat that presses down into the stream and so displaces the water around it. But every drop at the remotest bank is displaced according to its relations, and every drop on the bottom of the channel feels its part of the pressure.

One of the most decisive proofs of this great law of the perpetuity of our mental experiences is found in the familiar fact of *being turned round,* as it is called. We enter a strange place without having observed a turn we have made in our course. We have been going on a road leading northward, for example, and, without noticing it, we have turned into one leading eastward; this road will seem to us afterwards as if leading northward. The sun seems to us to rise in the south. We reason against the impression, but the first impression resists evidence and argument. If our intelligence is corrected, often our governing impulses follow the first impression whenever we are off our guard. If we are thus turned round in a strange city, we may move aright through one or two streets while we are guarding ourselves against being misled by the feeling; but as soon as we surrender our movements to the control of our mere purpose in walking, we turn north when we should go east. It is marvellous with what persistence such impressions in regard to the points of the compass abide in the mind. The writer has known of an instance when such an erroneous impression remained fresh and strong for many years, and, although the street was traversed several times a day, still remained so vigorous and strong as to require habitual care and watchfulness to prevent mistake.

From Extraordinary Experience. 4. Facts of extraordinary experience confirm in our minds the conviction that what is once experienced by the mind is ever retained by it.

In insanity, it is often observed that thoughts are recalled which, both before and after the attack, were beyond all power of recollection. These retained thoughts, also, reappear with a marvellous freshness and completeness. The records of Hospitals for the insane are replete with instances of mental activity stored with thoughts and feelings and volitions from past experience that have so outmeasured the seeming capacity of the mind in a sound state as to be well nigh incredible. A gentleman in an insane retreat, says Dr. Rush, astonished everybody with his displays of oratory; and a lady, he writes, sang hymns and songs of her own composition so perfect that he used to hang upon them with delight whenever he visited her, and yet she had never shown a talent for poetry or music in any previous part of her life.

In fever, also, similar facts are frequently occurring. The Countess de Laval was wont in sickness to talk in her sleep in a language that the servants could not understand. A nurse from her native province, Brittany, being engaged to attend her, however, recognized the strange speech as her native tongue. Yet when awake the Countess did not understand a word of Breton, so entirely had it seemingly passed from her recollection.

Coleridge narrates a similar case of an illiterate young woman of four or five and twenty, who in a nervous fever was heard to talk in Latin, Greek, and Hebrew. The matter excited great interest and on a protracted and thorough investigation it was ascertained that at the age of nine years she had been taken in charity into the house of a learned pastor where she remained some years until his death. This pastor had been accustomed to walk up and down a passage of his house into which the kitchen-door opened and to read aloud from his favorite books in these learned languages. Sheets full of her utterances were taken down from her lips; they had no connection with one another, yet each sentence was complete and coherent with itself. It was discovered thus that these recitations of her master, from languages utterly unknown to her, had been retained so perfectly that even after the lapse of years, in the excitement of the sensibility in fever, she was able to render them distinctly and perfectly.

The experiences of persons recovered when near being drowned are in evidence here. They frequently say that the events of their whole lives pass in clear, distinct, full review before them. A case narrated by the subject to the author, is a sufficient exemplification. He had been entrusted with the keeping of a package of valuable papers by a relative when about taking a long journey. On the return of his friend, he was utterly unable to recall where he had placed the package. The most diligent and careful search as well as every effort of recollection failed to discover the desired package. Years after when bathing, he was seized with cramp and sank. He rose and sank again; and, as he was just sinking the third time, a companion succeeded in reaching and rescuing him. During the momentary interval between his disappearance the third time and his being seized by his companion, his whole life in its minute incidents passed in review before his mind; and among them the fact of his secreting the package and the place where he had concealed it. He proceeded immediately to the spot, where he found, just as he had placed it, what he had so long sought in vain.

A singular case of catalepsy, cited by Hamilton from a German work by Abel, is also in evidence that men's forgetfulness is not decisive proof against this perpetuity of mental experiences. In this case a young man, some six minutes after falling asleep, would begin to speak distinctly and almost always of the same objects and connected events, so that he carried on from night to night the same history. On awakening he had no remembrance whatever of his dreaming thoughts. Thus it was that by day he was the poor apprentice of a merchant; while by night he was a married man, the father of a family, a senator, and in affluent circumstances. If during his vision, any thing was said to him in regard to what occurred to him during the waking state, he would declare that it was all a dream.

While memory proper has for its essential attribute this character of retentiveness, it must be borne in mind that it is the retentiveness of an active nature. It is not the retentiveness of a rock or of steel which retains the lines which have been inscribed upon them. It is not the retentiveness of a vessel or cell that retains what has been poured in or packed away in it. It is not the retentiveness of an animal organ that retains the disposition of fibres or of cells which it may in any way have received. It is the retentiveness of an enduring active being, which not only receives impressions according to its own active nature, but uses these impressions afterwards more or less in all its ceaseless action.

Memory is to be conceived of as something more than a mere capability to recall past experiences. At least an empty capability of recollection does not express the full truth. These past experiences live on in a true sense and are active parts of the present mental being. The man of learning, of achievement, of suffering, is more than a being capable of recalling his past thoughts and deeds and trials. These experiences have entered into his soul and have enlarged and strengthened it; whether any one or more of them are distinctly in his present consciousness or not, he is more and different because of them; his words, his steps, all he does, evinces a fulness of power, a mode and form of movement, a character in short altogether different from a nature that had not had these experiences. The adult man differs from the child in something more than a mere capability of bringing into his consciousness certain things of the past. His consciousness is a capability, a power indeed, but a capability, a power, replete with knowledge, with skill, with passion.

§ 122. This law of retentiveness in mind as an active nature imposes three conditions of a good memory. They are founded respectively in the subject-matter of remembrance—in what is to be remembered; in the relation of each thing remembered to other things in the mind; and in the character of mind itself.

Three conditions of good memory.

§ 123.   1. The first condition of a good memory is that it accepts as what it is to retain, so far as possible, only what the mind may need or wish to use.

1. Worthy objects.

The mind, as we have seen, is subject to impressions from without, beyond its control. It has consequently feelings, thoughts, and volitions, which it could not altogether prevent. But it has nevertheless a power to a large extent both to regulate the kind of impressions to which it will allow itself to be open, and still more to shape them when received to its own uses. Now, nothing can be more important to all the great ends of memory, which is to retain forever for future use and influence upon the mind every feeling and thought and desire and purpose, than that just the right impressions, the right feelings, the right thoughts, the right volitions, should enter the memory. No feeling or thought or intention which we do not feel willing to have ever confronting us, ever shaping and coloring our destiny, ever present in our soul's very being, and working in us and on us whether we are conscious of it or not, whether we are willing or not, should, if it lie in our power to prevent it, ever be allowed to enter our minds. If any such impression comes upon us, then should it be so controlled and shaped as that ever afterwards when it reappears it shall be in a welcome form, and shall when we are unconscious of it be silently influencing our whole mental action favorably. Our observations, our readings, our reflections, our reveries even, should be such as will fill our memories with nothing but what we shall in every moment of our subsequent lives be glad to find there. The scenes, the objects, the associates, the books, all the occasions of our feeling and acting should be carefully regulated with this view and under this momentous consideration, that what they bring into our minds is to remain in us perpetually.

Particularly does this characteristic of a good memory—good for the mind's uses—prescribe that our observations and our thoughts be accurate and true, as we would not have falsehood or error to mar all our coming thought.

It prescribes, also, that our feelings and acts should be in the most perfect form into which our imaginations can shape impressions or suggestions; that every recurring thought and imagination may shed the radiance of beauty on all our inward experience. A feeling of pain thus, that a stroke of malice has inflicted, may continue to exist in our minds to color more or less their whole future, according as our imaginations, reacting on the received impression, invests the pain in a form of forgiveness and of pity, or of bitter resentment. Thus it may be with all impressions which in themselves may be undesirable. They may be put in forms that shall never recur but to gladden and refresh us.

It prescribes, moreover, that all our intentions, our plans, our endeavors, and all other voluntary acts should be just and right, so that none shall in all the future of our being be present in our minds to disturb, to annoy, or to bring righteous retribution in evil of any kind upon us.

§ 124. 2. The second condition of a good memory is that it so links in every fresh experience with past acts and feelings, as to make it most easily to be recalled, and to work most serviceably for all that the mind can properly desire.

<small>2. Aid of association.</small>

The importance of observing this principle in the culture of the memory will be more fully seen when the nature and laws of association are explained. This will be the topic of the next chapter.

§ 125. 3. The third condition of a good memory is that it enlists a lively energy of the whole mind in its interest.

<small>3. Mental interest.</small>

What is to be preferably remembered, what is to be present with us when we may happen specially to need it, what is to influence greatly all our future thought and feeling, should receive the most of the mind's vigor and strength. What we receive listlessly, while it may in a sense abide with us, can influence us but little, can be little at our command in the time of need. What engages our interest deeply and vividly, we retain best for use and service.

§ 126. Under the great principle of memory that every act and feeling abides forever in the mind's active nature, in its degree and way shaping and coloring all its movements, we have thus the three specific rules of memory that have been stated:

<small>Rules of memory.</small>

1. That, so far as may be, only true thoughts, beautiful imaginings, good intentions and endeavors enter our memories.

2. That all fresh acts and feelings be properly associated with existing thoughts and feelings; and

3. That what we wish to be most ready and serviceable in our memories engage at the time the mind's utmost interest, attention, and care.

CHAPTER XIII.

MENTAL REPRODUCTION.

Mental reproduction. § 127. By MENTAL REPRODUCTION is meant the reawakening in the present consciousness of previous acts and feelings. It applies to all the phenomena of mind—to its feelings, its thoughts, and its volitions. It is the necessary condition of imagination. It has been also named suggestion and association of ideas.

Explained. In order to a right understanding of this function, it is necessary to bear in mind distinctly that the ideas reawakened by this reproductive function are introduced into a living nature, having at the time feelings, thoughts, and volitions, so that the old or reawakened ideas co-exist with the new or present, and with them form one complex condition of the mind—one complex body of mental experience.

Spontaneous or voluntary. § 128. Mental reproduction may be either spontaneous or voluntary; may take place in the unregulated flow of the mind's activity, or may be more or less directly controlled by the will.

Spontaneous reproduction takes place as the

characteristic element in what is known under the more familiar name of REVERY.

In revery the mind surrenders itself with no conscious control to its own current, so to speak, allowing thought and feeling to flow on according to their own tendency. In this state we discover, as we reproduce it for study into our thought, that one thought is followed by another, one feeling by another, and thought is followed by feeling, as well as feeling by thought. The interesting question arises, what determines this suggestion of one mental state by another. "Therein," says one, "lies the greatest mystery of all philosophy." This mystery psychologists have sought to explain by indicating the general principles or laws of reproduction, or of suggestion, otherwise called the laws of the *association of ideas*.

That there is some bond of connection, that there is some ground of association, psychologists have admitted or assumed. These thoughts and feelings that pass along through the mind one after another, they agree, do not come hap-hazard; they succeed one another under a governing law.

It may be remarked here that beyond all reasonable question the succession of thoughts and feelings in dreams and in insanity, is similar to the succession in revery, and with some modifications is subject to the same laws.

Laws of mental association. From the earliest times philosophers have presented, one after another, each his own enumeration of the laws of association. Sir William Hamilton has gathered up these proposed principles and reduced them all to the following classes. Thoughts are associated, he says, in the respective opinions of these philosophers, 1, if connected in time; 2, if adjoining in space; 3, if related as cause and effect, as means and ends, or as whole and part; 4, if similar or in contrast; 5, if products of the same mental power, or of different powers conversant with the same object; 6, if the objects of the thoughts are the sign and the signified; 7, if their objects are directed by the same word or sound. He himself thinks these principles may all be reduced under one law, which he calls the law of Redintegration, (restoration to a whole), and which he thus enounces: "Those thoughts suggest each other which had previously constituted parts of the same entire or total act of cognition."

The law as thus enounced, it must be said however, is palpably insufficient to meet the demands of the problem. It does not embrace feelings or volitions; no explanation whatever is given of the fact that one feeling draws in another feeling, and one purpose another purpose, nor of the fact that feelings suggest thoughts. Nor does it even cover the familiar fact that a perfectly new thought, which therefore could not have previously constituted a part of any act of cognition, suggests old thoughts or new thoughts. I meet a stranger in the street, whom I have never seen or heard of before; the sight may suggest any one of ten thousand different thoughts or feelings.

The same fatal deficiency in meeting the demands of the problem, characterizes other attempts to gather up into an exhaustive statement the manifold grounds of association or suggestion. It is true that one part of a past thought may suggest another part; it is true also that some similarity in thoughts is a bond which unites them so that they may suggest one another; it is true that connection in time or space, or as cause and effect, is a ground of suggestion; and so of all the other proposed laws; they are grounds, but all together they do not make up all the grounds of suggestion. The problem to be solved, the mystery to be explained, is analogous to this. A particle of the green mud from the Ohio is found united in the Great River with a particle of the red mud from the Arkansas; they come together under the operation of inflexible laws of nature; can now—this is the problem—can these laws be stated and be traced in their operation to their bringing together these two particles? The analogy would be more exact if we were to suppose all the particles that have ever come into the channel of the Great River to be brought to a stand against some immense perpendicular barrier, and the river under its own laws to be shifting continually the positions of the entire mass of particles and thus bringing the two particles into ever new yet ever shifting positions and relations. That the two particles meet and unite is undoubtedly due to some fixed law or laws of nature. We have the great law of gravity bearing the two down together in the same open channel; we have the probability that if the two particles entered the same part of the current at the same time, they might come together. If they had been subject to equivalent forces of repulsion from the banks, of impulse from winds, of depression from floating objects, of rarefaction from heat, and the like, we have in these conditions other reasons for their being together. But so manifold are the influences at work, that human reason recoils from the task of tracing them all.

<small>General principle.</small> It is so with the associations of any two thoughts or feelings in the mind. The one principle that covers the whole matter is simply this: They are states of the same one mind, as the two particles supposed are parts of the same rolling river; and this mind has power, under favoring conditions, to call forth into consciousness, within certain limits at least, any part of its collected activity of thought and feeling and volitions; and therefore power within such limits to connect any present state of consciousness with such recollected thought or feeling or volition, and so bring to the surface of its great volume of accumulated experiences, that is into distinct consciousness, a new mental experience. It is not presumable that any absolutely universal law of association can be framed other than this, that all associated ideas must belong to the one same mind; and that any one idea may, in the possibility of things, be associated with any other idea of the same mind; just as two particles of white and red mud in the Great River must, to be brought together, be in the same stream, and any two in that stream

may, in the possibility of things, be brought together.

But in co-existence with this general law there may be, and in fact there are, other more specific laws implying the existence of specific causes which may effect the association of ideas. As these more specific laws may be convenient helps to recollection, it may be of service to make a formal and collective statement of the principles of association. Whatever limitations of this power of recollection may exist, it may be remarked, pertain only to the mind as finite ; not to the relation between any two thoughts or feelings. The general principle is, that nothing but the weakness of mind as a finite nature hinders the association of any two mental acts or feelings which the mind has ever experienced. The principle implies both that no mental exercise ever becomes annihilated so that on this account it cannot be recalled, and also that every exercise is so connected with every other that the one may possibly suggest the other.

§ 129. LAWS OF MENTAL ASSOCIATION. 1. Any part of the mind's total experience may be associated with any other, and so in favoring conditions suggest it. In briefer terms: In the same mind any idea may suggest any other idea.

Special laws. 1. Any part of experience may be associated with any other.

This is the comprehensive law. It includes all kinds of mental experience, feelings and volitions as well as thoughts. Any feeling may suggest any other feeling, or any thought, or any volition which has entered into the mind's experience. By suggesting here, it should be borne in mind, is meant bringing forth from unconscious experience into distinct consciousness.

§ 130. 2. Any part of the mind's experience may suggest any co-ordinate part :—any idea suggests with special power a co-ordinate idea.

2. Co-ordinate parts.

A feeling may suggest a co-ordinate feeling. A man in a mood of excited feeling is easily drawn into another feeling. We pass more easily to weeping from laughter than from an utterly unfeeling state.

In the same way thought helps thought. It is a common practice with intellectual men to prepare themselves for clear, accurate, vigorous thought on any subject by putting themselves on the intense study of some other subject into which the mind can more readily enter. Lord Brougham trained himself for a great intellectual effort by a long and intense study of Demosthenes' Oration on the Crown.

An active will in any one direction easily slides into action in any other direction. It is easier thus to enlist an active man in a new enterprise than the dull and idle.

§ 131. 3. A generic part of mental experience may suggest any subordinate part ; and conversely the subordinate may suggest the generic or comprehensive. Ideas that are respectively super-ordinate and subordinate to each other mutually suggest each other.

3. Subordinate parts.

A man in an angry mood easily breaks out in

new passion towards any particular object, whether newly presented or reawakened in memory. Compassion towards a single sufferer inclines to pity for all of the class, for general good will.

To recall the individual of a class to our thought, we naturally turn to the class and from that seek to recall the desired object; or conversely, having the individual in our mind and desiring to recall the class, we naturally endeavor to realize our wish by thinking of the individual.

It is the same with the will. We form a general purpose; it brings on all subordinate purposes. We resolve to speak, and the determination leads on to an indefinite number of subordinate purposes controlling our attitude, our gesticulation, our sentences, our respiration, our vocalization, our single words, our articulations. The single purpose reacts too on the general purpose and carries it on, keeps it alive, as well as guides and modifies it. Nothing better seems to revive a dormant resolution than to do some particular thing involved in it, or which may be made part of it.

4. Association by objects.
§ 132. 4. Parts of the same object of mental activity suggest co-ordinate or subordinate parts.

This is but another form in fact of stating the preceding laws; it respects directly the object of mental action, while they respect the mental action itself.

5. By symbols.
§ 133. 5. Parts of the same symbols or signs of objects in the same way suggest other co-ordinate or subordinate parts.

If the mind has before it either part of the word, *farewell, fare*, or *well*, that part may suggest the other; or it may suggest any one of the parts of which it is composed. The philologist, for instance, may think of one or another of the sounds or the written characters which constitute the word. The cherubs in Raphael's Sistine Madonna will suggest the Madonna herself or any other part of the picture, or any posture, expression, or feature in the cherubs themselves.

§ 134. 6. Mental experiences of more recent occurrence have greater suggestive power: The more recent the idea, the greater is its power to suggest.

6. The more recent.

This law of association, it will be observed, is of a different source and character from the preceding. It is founded in the attribute of growth that we have found to belong to the human mind. Every new stage of its existence brings in a new stage of growth, a fresh life, a larger development. Such at least is the general law. The most recent life consequently has a greater vigor and intensity.

This fact of association we all familiarly recognize. We recall the occurrences of yesterday more readily than those of the last year; and these more readily than those of ten years before. The law, of course, regards experiences of the same character otherwise, such as experiences of the same closeness of connection with the suggesting act or feeling; or experiences of the same interest and importance.

An apparent exception to this law is found in the

experience of aged persons, who often recall the events of childhood and youth more readily and more vividly than those of later years. But this fact may be accounted for in part at least, on the ground that their habitual thoughts at this period of life run in the channels of earlier experiences. These, therefore, from their being revived and lived over again, are really the freshest and latest in their minds. Farther than this, other principles of association may come in. External scenes and objects, individual associations, and numberless influences from personal attachments and repulsions, come in to make parts of a mental experience by which other parts are suggested.

7. Intensity of experience. § 135. VII. The intensity of the mental experience is an important element in association or suggestion :— The more vivid the idea, the stronger is its suggestive power.

Intense feeling kindles at once from the faintest impression. An angry man bursts into stronger passion from a provocation of any kind. Energetic thinking fuses all the particular thoughts together, so that as if inseparable one cannot return into the mind without drawing in the others. Our resolutions carry all subordinate purposes just in proportion as they are strong and energetic, enlisting the whole soul. When such a governing purpose is earnest and decided, all purposes that are foreign to it, even if occasion should suggest them, give way at once. When, likewise, a specific purpose is thus earnest, all other specific purposes under the same general resolution, fall in more easily. Weak souls are ever characterized as inconsistent.

If the demand be pushed farther for the reason why in any particular case this part is suggested rather than that, while sometimes a more subordinate law may be assigned, ultimately we are obliged from the finiteness of our power to fall back upon the first general law given,—the unity of the mind itself carrying in its complex activity all the special activities of feeling, of thought, and of volition, just as we are forced, in attempting to account for the union of the two particles of mud to fall back on the general fact of their being in the same whirling, rolling stream. So many forces come in of such various intensity, from the world without, from the state of the body and its nervous organism; from the habits, tastes, moods, of the individual mind itself; that it is beyond the power of created intelligence fully to account for all the associations of ideas that it experiences. It must be recollected that these forces come up as well from the vast volume of our unconscious experience as from the mere surface of mental action which our distinct consciousness takes up.

2. Recollection. § 136. Mental reproduction may be to some extent under the direction of the will. Voluntary reproduction is familiarly denoted by the term *recollection*.

We recognize this fact that reproduction is in some measure subject to our wills when in our desire to recall some past experience we endeavor to direct our thoughts or feeling towards it. We do this in two different ways: positively by keeping in our consciousness some experience associated with what we wish to recall; and negatively, by repelling thoughts and feelings that are more foreign to it.

The positive endeavor to recall a past experience will of course best be guided by the laws of association. It assumes some feeling, or thought, or volition from which it is to proceed as its necessary ground and starting point. With this experience in the consciousness, recollection properly sets out and then puts itself under the lead of the laws of association as they have been already stated. The best rules of recollection may accordingly be thus summarily given:—

§ 137. RULES OF RECOLLECTION. I. Recall feeling by feeling, thought by thought, purpose by purpose.

1. Like department of mind.

Early affection for a friend long separated from us may best be revived from a similar state of affection in exercise towards a friend still with us. In like manner a former thought is best revived when thinking rather than feeling or endeavoring is the predominant characteristic of the mind. Free action in the same way revives a dormant purpose or endeavor. Even if the mind in a state of excited feeling desires to recall the train of thought out of which the feeling rose or with which it was associated, for the most part success will be most probable if the existing feeling first recall the old feeling and then that feeling revive its associated thought.

§ 138. II. Recall ideas through the relation of whole and part.

e. Whole and part.

If the feeling or thought or purpose to be recalled be generic or comprehensive, start from a subordinate experience; if subordinate, start from a generic or comprehensive experience.

To revive a governing disposition of filial dutifulness, a present purpose in doing some particular act of filial duty will be the most hopeful. So a general thought is best recalled by thinking of some particular fact or instance in which that principle is exemplified. As for example, in recalling the general law of material gravitation, I may succeed best by beginning with the law as instanced in a falling weight and thinking of the number of feet of fall in the first second, the number in the second, the number in the third, and so on.

So to recall a subordinate purpose, it is best, if it be practicable, to begin with a generic or governing endeavor. To revive a neglected religious duty, the most hopeful method is to begin with a freshened endeavor to do all religious duty. To recall a specific thought, it is well to begin with the general law that comprises that thought.

§ 139. III. Recall objects through the same relation of whole and part, as associated either with one another or with the mental state which they respect.

3. By association.

§ 140. IV. Words and other symbols are most suggestive of like words and symbols, or of the objects or mental states with which they are associated.

4. By symbols.

CHAPTER XIV.

THE ARTISTIC, PHILOSOPHICAL, AND PRACTICAL IMAGINATION.

Manifold functions of the imagination.
§ 141. Ideals, as the proper products of the imagination, may be distinguished into three general classes, corresponding to the three general functions of the mind, feeling, thinking, willing; also to the three generic objects of all mental activity, the beautiful, the true, and the good.

We have thus three functions of the imagination determined in reference to the character of its product or ideal: three forms of the imagination as an active power:—

1. *The Artistic Imagination.*
2. *The Philosophical Imagination.*
3. *The Practical Imagination.*

It must be borne in mind here as every where, that these products of the imagination, these ideals, are so distinguished only as they are more prominently characterized respectively either as beautiful, true, or good. Every act of mind, every idea, has necessarily each of these attributes in some degree; but it may have one more prominent than the others which thus gives character to the act. If an artist frames an ideal of a virtue, as, for instance, of patriotism, or of filial affection, he necessarily regards more or less the principles of truth, of intelligence, and also those of right-doing. But his governing end being a beautiful form, his ideal is characterized as properly artistic, not philosophical nor practical. The philosopher, in the same way, although his governing end is truth, and his labor is to attain or set forth what is true, still must regard the form which his speculations take and the effect in some way or other which they may work. But his prominent ideal being the true, it is easily distinguished by this characteristic; it differs from a mere ideal to be marked by its beauty. A geometrical treatise does not properly take on a poetical form. The practical man, moreover, cannot disregard the form of his product, nor the essential attributes—the truth—of things; but his act is characteristically distinct from the proper work of the artist and of the philosopher.

Still further, the degrees in which the one or the other of these three great attributes of all mental activity, the attributes of form, truth, and practical effect, predominate in ideals, vary indefinitely. The practical philanthropist, who aims to do good as his chief governing aim, may put his act of kindness into such a frame of loveliness that we hesitate which to admire most, the beauty or the goodness of his act. In truth the imagination which shaped his act may be regarded as having been both artistic and practical; both graceful and beneficent. It may have been also eminently wise, conformed in all particulars to the truth of things. His act will be characterized as good, or beautiful, or true, according as one or another of these attributes is recognized as predominant in it.

§ 142. THE ARTISTIC IMAGINATION produces ideals characterized by their form, as beautiful or the opposite.

1. Artistic imagination.

The governing end in the artistic imagination is form. The work may be more or less conformed to truth, may more or less promote truth; it may proceed from a general benevolent intention and may be productive of good; but the artist in his own proper specific work, looks to the form of his product. His work will indeed be more or less perfect in form according as he more or less strictly conforms his work to the truth of things, or as he works more or less perfectly in the line of goodness; yet we have no difficulty in recognizing the work as characteristically a piece of art and not a work of speculation or of morality.

It is the proper province of the science of æsthetics to ascertain and apply the laws of the artistic imagination both in the production and in the interpretation of beauty.

§ 143. THE PHILOSOPHICAL IMAGINATION produces ideals characterized by their essence as true or the opposite.

2. Philosophical.

The philosophical imagination seeks truth as the governing end of its activity. The artistic imagination produces for the form's sake, although not transgressing the laws of the true; the philosophical imagination, on the other hand, produces for the truth's sake, although not transgressing the laws of the beautiful. The mental act has a twofold aspect; one and single in itself; it yet engages the imagination or the faculty of form and the intelligence as the faculty of the true. If we regard the mental activity on the side of the imagination we denominate it the philosophical imagination; if we regard it on the side of the intelligence, we call it the intellectual representation, § 188.

It is the proper province of the science of logic to expound the laws by which the philosophical imagination or the faculty of the intelligence acting in the representation of truth or knowledge, is to be governed. This science thus determines the valid forms of all thought or knowledge.

3. Practical. § 144 THE PRACTICAL IMAGINATION produces ideals characterized by their tendency to a result or effect as good or the opposite.

The practical imagination frames ideals of character to which the whole activity of the soul is to be shaped. It devises plans of active exertion and methods of execution. As the life of the artist is characteristically that of one who is ever shaping beautiful forms, idealizing for the purpose of impressing beauty, and as the life of the student and the philosopher is characteristically a life busy with framing new and truer ideas of doctrine, of objects, of events, so the life of the practical man is characteristically the life of one busy in devising schemes of exertion, new pursuits, new enterprises, new methods of operation.

It is the proper province of the science of ethics, in its broadest sense as comprising not only the duties of religion and morality, but also the acts ol social life, of polity, civil and domestic economy, and those which pertain to personal well-being, to bodily and mental health and vigor, as well as the fulfilment of man's destiny as an active being,—it is the province of this broad science to unfold the laws by which practice in all these departments is to be regulated and controlled.

BOOK III.

THE INTELLIGENCE.

CHAPTER I.

ITS NATURE AND MODIFICATIONS.

The intelligence defined.
§ 147. THE INTELLIGENCE IS THE MIND'S FACULTY OF KNOWING. This faculty is otherwise known as the Cognitive Faculty, or the Faculty of Cognition, and also as the Intellect. Its function is simply that of knowing; and all knowing is by this faculty alone.

Its modifications.
§ 148. The intelligence is diversely modified in various respects.

It is modified, first, in respect to its completeness or incompleteness.

In respect to completeness.
A complete knowledge, a complete cognition is attained, only when there is a subject united with some attribute—only, in other words, when there is an assertion

either affirmative or negative expressed or implied. There is a kind of knowledge which must be deemed to be in a certain sense incomplete which is preparatory and conditional to this complete knowledge, and which is attained when its object is apprehended simply, and prior to any distinguishing of it into subject and attribute.

In other words, an act of knowledge is conveniently for study distinguished into two stages, the preparatory stage and the completed stage.

Presentative knowledge. The preparatory stage of knowledge has been called PRESENTATIVE KNOWLEDGE.

As finite and dependent, the human mind can know nothing but as an object is given—is presented to it. The mind's reception of such presented object—its apprehension of such object —is a presentative knowledge.

Representative knowledge. But the intelligence cannot rest satisfied with this mere presentative knowledge. Its essential activity prompts to a further stage in which the object presented to it in the preceding stage is recognized in a two-fold aspect—as subject and as attribute. Thus if any object is given to it, as for instance *the sun*, the intelligence at once proceeds to apprehend it as having an attribute—*brightness ;* and its knowledge is complete only when the mind is in that state which is properly expressed in a proposition, *the sun is bright; sun* and *bright* are not two different things in reality, but it is the native function of the mind at once in every single object presented to it to recognize such object under this form—the form of a subject and an attribute which it unites. This complete knowledge is properly termed *representative knowledge*, as it implies, in addition to the first stage or presentative knowledge, a reflex act of the mind on the object presented to it.

In the earlier stages of mental development presentative knowledge has a greater relative prominence than in maturer life. The child perceives, simply apprehends, relatively more; but as he advances his simple apprehensions pass more habitually into proper reflection or judging.

2. In respect to object.   § 149. The intelligence, further, is modified in respect to the diverse character of its object.

We have found that the comprehensive object for the intelligence is *the true*. But *the true* ever embraces three distinct elements or constituents— the subject, the attribute, and the uniting element called the copula. These elements may severally vary in manifold ways. They so far modify the act of the intelligence in knowing.

In respect to sources of knowledge.   § 150. The intelligence, once more, is modified in respect to the sources of its knowledge.

Its objects are presented to it from two different directions, which it is very important to recognize distinctly. These objects are brought to it in part from without and presented to it through the external senses. The presentative knowledge thus attained is called a *perception* or

Perception.   *perceptive knowledge*. These objects are brought to it in part, moreover, from

Intuition. the mind itself—from its own phenomena. The presentative knowledge thus attained is called an *intuition* or *intuitive knowledge*.

4. In respect to other functions. § 151. The intelligence, finally, is modified in respect to the different functions of the mind itself.

These functions have already been recognized as three-fold—the sensibility, the intelligence, and the will—as the functions of the same indivisible nature, which in no exercise of any one function ever drops entirely either of its other functions. Every act of the intelligence is more or less modified by the sensibility and the will. But the intelligence sometimes acts upon its own operations. It takes knowledge of them. It is then said to be *conscious* of its states.

Not only this, but the human mind being both passive and active in every state, we have ever two sides to study—the passive side of knowledge, in which the intelligence apprehends its object, and the active side, in which the intelligence puts forth its object.

We shall, in the future exposition of the intelligence, treat in separate chapters of the generic forms of these several modifications.

CHAPTER II.

PERCEPTION.

Perception defined.

§ 152. PERCEPTION is that function of the intelligence by which an object presented through the senses is simply apprehended.

Here, as elsewhere, it is important to distinguish carefully the three-fold meaning of terms applied to the phenomena of the mind. The term *perception* is used to denote the faculty of perceiving; the exercise of this faculty, or the act of perceiving; and also the result of this act. The term *percept* has been proposed to denote the result or the product of perception.

An active function.

§ 153. Perception is the active or knowing side, sensation the passive or feeling side, of the same mental state.

We have already recognized the truth that the mind is in every experience both passive and active. This law of mind is formally proposed by Sir William Hamilton in its general form, as applied to all mental phenomena; it is specifically recognized by him in its application here in the summary statement: "Cognition and feeling are always co-existent." 1

perceive an orange at the same time that I have a sensation of it through the eye, the touch, the smell, or the taste.

Distinguished from sensation But while perception and sensation are but opposite sides of the same mental state, which has ever an active and a passive side, they are to be distinguished from each other in several important respects.

1. Sensation is the ground or occasion of the perception. It is, therefore, properly regarded as the logical antecedent of perception, and in this sense as prior to it.

2. Sensation is not only the ground of perception—not only conditions it so that perception cannot be without sensation—but it also determines and shapes perception. Only as perception conforms itself exactly to the sensation is it legitimate or sound.

§ 154. Generally and loosely speaking, *In inverse ratio to it.* sensation and perception are in the inverse ratio to each other. The stronger the sensation, the weaker the perception, and the stronger the perception, the weaker the sensation.

Sir William Hamilton has exemplified this general law in the comparison of the several special senses. In sight, perception is at the maximum, sensation at the minimum. We are hardly conscious of any feeling in seeing an ordinary object; we are conscious of a decided knowledge of objects that we see. We look at the orange; the sight of itself is without any feeling intense enough to be noticed; the knowledge of its being before us, of its being round and yellow, is perfect beyond that given by all the other senses combined. In hearing, there is far more of feeling than in sight; far less of knowledge. In taste and smell and special touch, feeling greatly predominates and the perception is relatively slight and limited.

If we take again any particular sense and regard it separately from the other senses, we notice that generally if the feeling is strong, perception is weak and the reverse. If the sensation of sight, for instance, be strong, we are dazzled—we feel intensely; but we perceive comparatively little.

<small>This law not absolute.</small> This law, however, cannot be adopted as absolute or universal. The sensation may be so weak as to occasion no perception at all, when by the law it should be at its maximum. The strength of the perception often varies directly, not inversely as the sensation. If a man touch me gently with his finger, I hardly feel it perhaps, and hardly perceive the fact that I am touched, or what touched me; I have but little knowledge because I have but little feeling. If he strike me violently with his cane I both feel and perceive intensely.

This general truth, however, is ever to be borne in mind that whatever the relation between the sensation and the perception in respect to their comparative intensity or strength, either one may become the object of consciousness to the exclusion of the other. The light may come streaming in from every visible object upon my eye and engage my whole mind with the mere feeling of its cheering impressions, so that I shall distinguish not a single object and have no conscious perception. Or I may so attend to the knowledge of particular objects as not to be distinctly aware of any sensation.

<small>Sphere of perception.</small> § 155. THE SPHERE OF PERCEPTION is the world of sensible objects—the entire realm of external phenomena of which we can have any intelligence.

<small>A presentative knowledge.</small> § 156. Perception is an act of presentative knowledge. It gives the knowledge of the object simply, without distinguishing it into subject and attribute.

I perceive an orange; but in the perception itself I only know it as an object without passing on to think it to be round or yellow.

<small>Immediate.</small> § 157. Perception gives accordingly only an immediate knowledge—a knowledge not mediated through the distinction of subject and attribute.

It is true that every object that can be known must have an attribute. It is true that the mind tends to pass beyond the stage of incomplete knowing to the complete knowledge under or through an attribute. But perception is confined to the first stage. It does not discriminate attributes. This discriminative or completed knowledge will be investigated hereafter.

CHAPTER III.

INTUITION.

Intuition defined.
§ 158. INTUITION is that function of the intelligence by which an object presented by the mind itself is apprehended.

We have distinguished sense-ideals from spiritual ideals. The same distinction exists between a product of perception—a percept—and a product or result of intuition. An object of perception is known through the external senses—through a sensation; an object of intuition is known through the mind's own action simply. The former belongs to the external world—the world of matter; the latter to the internal world—the world of mind.

Synonyms.
§ 159. Intuition has been variously designated as self-consciousness; the faculty of internal perception; and the faculty of internal apprehension.

The term *intuition* has been much used in a narrower sense to denote only the faculty of apprehending what are called self-evident or necessary truths. Intuitions, thus, in this narrower use, are necessary truths or ideas. The term *intuition* is also used by some writers in a larger sense to include all presentative knowledge, and consequently perception as well as apprehension of internal phenomena. In German literature it is commonly used in this large sense.

Shunning both of these opposite extremes in the use of the word, we shall employ it to denote that function of presentative knowledge by which the phenomena of mind are apprehended. Intuition and Perception thus constitute the total function of presentative knowledge—the former apprehending internal, the latter external phenomena.

Sphere.
§ 160. The sphere of Intuition is exactly defined as the sphere of mental phenomena. Of these we have recognized three general departments—feelings, cognitions, volitions.

If we have a sensation of an orange—a taste of it as sweet or juicy—we may apprehend that sensation in our intelligence; we may know the sensation itself as truly as we know in perception the orange itself that caused it. If we have a feeling of anger, we may know this feeling. This knowledge of the feeling is an intuition. Just so if we have a perception, we may know that perception; or if we have a volition, a purpose, we may know that act of will.

A presentative knowledge.
§ 161. An Intuition is an act of presentative knowledge. It presents the object simply—the feeling, the cognition, the volition.

Intuition is, therefore, an incomplete knowledge. It does not distinguish the feeling into some thing having an attribute. We have in an intuition only the knowledge of the feeling before recognizing that it is strong or weak, that it is real or imaginary. Our minds by the tendency of their nature press on to a complete knowledge. But it is convenient to recognize this completed knowledge as attained by two distinguished stages. Intuition brings us only over the first stage. It gives only incomplete and preparatory knowledge.

§ 162. An Intuition gives, accordingly, an immediate knowledge, in the sense that the knowledge it gives is not mediated to us through an attribution.

Immediate.

It follows from this that inasmuch as attribution ever gives a truth, an intuition properly regards an object, not a truth. If a truth, that is, if a proposition, be regarded in intuition, it is as an object simply; in the intuition there is no affirmation by the mind itself that the proposition is a true one.

CHAPTER IV.

THOUGHT.

Thought defined. § 163. THOUGHT is that function of the intelligence by which an object is known by means of an attribute.

The term *thought*, like intuition, is used in the threefold sense of (1) the faculty, (2) the exercise of the faculty, (3) the result or product of the exercise.

Synonyms. The faculty itself is, moreover, called by different names, as, the Discursive Faculty, the Elaborative Faculty, the Comparative Faculty, the Faculty of Relations.

Exemplified. The nature of thought may be thus exemplified. If an orange is presented to my sight or touch, I have a sensation and a perception of it. So far as I am only perceiving, I do not distinguish any attribute from the subject or that to which the attribute is supposed to belong; the perception does not reach the distinction expressed in the proposition, the orange is round. It matters not whether it be apprehended by me as an object or as an attribute, as a round thing or as roundness and yellowness. Perception carries my mind only through the first or preparatory stage of knowing. But my mind passes on to the second stage or that of a completed knowledge, and then it has a *thought* of the orange, which is properly and fully expressed in the proposition, *this thing is round*. I have now (1) a subject of which an attribute is thought—*this thing*, (2) an attribute belonging to this subject—round, and (3) that which is expressed by the word *is*, which identifies this subject and this attribute as one and the same. This is a typical form of all primitive thought, to which all thought however complex, however derived, may ever be referred back as the standard and model of all. I think when I distinguish in a perception or intuition attribute and subject, so that I can affirm the attribute of the subject as in the proposition: *the orange is round*.

It will be observed that the thought, *this thing is round*, is before all proper abstraction, before all analysis, before all generalization. A blind person for the first time coming into the warm rays of the sun might have a thought of a thing as warm without knowing anything else about the sun. If his mind were left to its own tendency he could, on perceiving the warmth, proceed to a completed knowledge by thinking a subject as having an attribute. He would have the thought: *the sun is warm*. But in this he would not have abstracted any thing, any attribute from any other thing or from any other attribute, for there had been given him but one thing, one attribute. He had not analyzed any thing or any attribute; for the thing was one and single and the attribute was one and simple, and neither therefore could be analyzed. He had not generalized; for this thing might have been to him the first and only thing of a class of warm things; it might have been to him the first conscious experience of the attribute of warmth. He could not of course in this case give the sun or warmth a name; for he had only and for the first time the thought, and naming, language, must necessarily follow, not precede thought. Abstraction, analysis, generalization, are processes which are applicable properly to complex and to derivative thought and apply to aggregated subjects and attributes; single and simple thought may take place without any of these processes. In order to obtain a clear and accurate notion of thought in its essential nature it is desirable to clear our view from all those processes which are not of the very essence of thought; from all those processes accordingly which can be applied only to complex or derivative thought.

§ 164. Thought follows and pre-supposes perception and intuition, the one or the other, as perception follows and pre-supposes sensation, and intuition follows and pre-supposes internal experience.

Order of experience.

The progress of the mind from the perception to the thought may be more or less rapid. It may be instantaneous as is the transition from sensation to perception, or the mind may linger on the perception to obtain a deeper and fuller impression; and thus it may happen that the progress may be arrested and the perception never ripen into ful thought.

It is to be remarked, also, that a previous thought as well as a perception or intuition may be the antecedent to a new thought. The finiteness and dependence of the human mind, however, compel us to the belief that perception, perhaps intuition also, must have preceded the first thought.

§ 165. The three essential elements of every thought are :—

Elements of thought.

1. THE SUBJECT, or that of which some attribute is thought;

2. THE ATTRIBUTE, or that which is thought of the subject; and

3. The COPULA, or that which unites or identifies the subject and the attribute.

These three elements are essential in all valid thought, whether primitive or derivative. If not expressed they must be implied; explicitly or implicitly, they exist in all legitimate thought. There is ever to be found a subject implying an attribute belonging to it, an attribute implying a subject to which it belongs, and the union or identification of this subject and this attribute. In the thought *this thing is round* the subject, *this thing*—orange—is not really different from the attribute; we do not apprehend *the orange*, and then *roundness;* it is the same as the attribute, and is in fact identified with it in the thought by the copula, *is.*

Subject and attribute are accordingly correlatives; the one necessarily implies the other; neither can be without the other. They denote distinctions which exist only in thought, not in the reality of things. When we speak of the subject as the unknown basis of attributes, we can mean only that of subject apart from its correlative attribute, we can know nothing; we know nothing except through some attribute and can know no attribute except as we know so far at least some subject to which it belongs. Knowledge is in fact when full and complete, nothing but the recognition of an object as something with an attribute.

The term *substance* is synonymous with *subject;* as is also *substratum*. They are all words from the Latin and alike point to that which underlies attributes. *Substance* and *substratum* are used more in metaphysical discourse, while *subject* is a technical word used in logical science, although not confined to this use. An attribute expressed in a proposition, is in logic termed a *predicate*. The subject and predicate in a logical proposition are called *terms*, from the Latin *termini*, limits, being the terminal elements, while the copula is the middle and connecting element of the proposition.

§ 166. Thought is properly called the *discursive* intelligence, inasmuch as, when a perception is presented to it, the mind in thought runs in two directions—*discurrit*—recognizing the single object presented in the perception under the twofold form of subject and attribute.

<small>Thought discursive.</small>

The object remains the same; it is still single. The change from the singleness in the perception to the twofoldness in the thought is in the mind alone. But the mind retains the original singleness in the object by its identifying the twofold members of the thought through the copula.

Forms. § 167. There are three generic forms of thought regarded as product: *the judgment; the concept; the reasoning.*

The judgment is the primitive form of thought; the concept and the reasoning are derivative forms from the judgment.

§ 168. THE JUDGMENT may be defined
1. The judgment. as that form of thought in which an object is identified with an attribute.

The regular form of a judgment in words is *the proposition*, which may be defined to be *a judgment expressed in words.*

In Grammar the proposition is known as the *sentence.*

§ 169. THE CONCEPT is a derivative from
2. The concept. the judgment. It is formed from the matter of the judgment—from its terms—and may be derived either from the subject or from the predicate.

The concept ever pre-supposes a judgment, and can be validated as a sound product of thought only by being referred to the judgment from which it is derived. Its name, from the Latin *con-ceptum*, imports that it is from its very nature *taken with* the other term of the judgment from which it comes. It is also called *conception*, which term is likewise used to denote the act of the intelligence in forming the concept and also the intelligence itself when exercising this function of conceiving.

If it come from the subject it is a subject concept; if from the predicate or attribute, it is a predicate concept. These two kinds of concept have characteristics widely diverse. Not a little confusion and error arise from a failure to distinguish them

Generalization. If we unite the subjects of judgments having the same predicate, we have *generic* or *class-concepts*, which are expressed in grammatical *class-nouns*. The process of thought in deriving this kind of concepts is *Generalization*.

Determination. If we unite the predicates of several judgments, we have *composite concepts*, expressed in grammatical *abstracts*. This is the logical process of *Determination*.

Category. We may think of an attribute as a subject. We may take thus the attribute *round* and think of it as having this or that attribute. We express it in that case in the form of a noun—*roundness;* as we say, *the roundness is perfect, the roundness is imperfect; it is that of a circle or that of a sphere;* and the like. Then we may unite several subjects of this kind when parts of judgments having the same predicate, and we have a class of attributes. A class of attributes is called in distinction from a class of origina. subjects, a *category*, from a Greek word signifying *predicated*. A CATEGORY is a class of attributes.

3. The reasoning. § 170. THE REASONING is derivative from a judgment, but appears still in

the form of a judgment and may be expressed in language by a proposition; while a concept is expressed in a grammatical noun. It may be defined as a derived judgment.

The derivation may consist in a change of the form or in a change of the matter, of the primitive judgment.

The derivation may be by a single step as in *immediate reasonings*; or by two or more steps as in *mediate reasonings*.

It is the province of logic to unfold the laws of thought and the different forms of valid thought, distinguishing the different kinds, with their several characteristics.

Returning now to the type-form of a primitive thought—of a judgment, as *the orange is round*—we observe that the attribute *round* is contained within the subject, *the orange*. But there are attributes which lie without the subject. We may attribute to the orange that it is *in the hand* that presents it to us, that it is *now* before me; that it is *one of a number*, and the like. Some attributes accordingly are intrinsic, others extrinsic to the subjects; obviously there can be no other attributes conceivable.

Attributes. § 171. All attributes are accordingly distinguished into two classes, one *Intrinsic*, the other *Extrinsic*, to the subject. INTRINSIC ATTRIBUTES are those which

1. Intrinsic lie wholly within the subject and may be thought when that alone is presented to the mind; as *round, yellow, sweet, juicy*.

2. Extrins.c.

EXTRINSIC ATTRIBUTES are those which lie without the subject and are thought only when something besides the subject is presented to the mind; as *in the hand, present, one of a class.*

Attributes of quality and of action.

§ 172. INTRINSIC ATTRIBUTES are of two species; *Qualities* and *Actions;* the former regarding the subject as at rest, as simply being; the latter regarding it as an activity, as acting. *Round, yellow,* thus are attributes of quality; *nourishing, cooling, fermenting,* are attributes of action. The latter imply, while the former do not imply, an object.

Properties.

§ 173. Intrinsic Attributes, also called Properties, moreover, are distinguished into the two species of *Essential* and *Accidental,* the former being necessary to the being of the subject; the latter not thus necessary. *Round, yellow,* or at least some attributes of figure and of color, are essential to the orange; *specked, decaying,* are not thus essential.

Attributes of relation and condition.

§ 174. Extrinsic Attributes are distributed into the species of attributes of *Relation* and attributes of *Condition.* *One of a class, larger than the others,* are attributes of relation: *present, now,* are attributes of condition.

CHAPTER V.

THE CATEGORIES OF THOUGHT.

Category explained.
§ 175. By a Category is understood, as before stated, § 169, a class of attributes, as distinguished from a class of things or of subjects. In the study of the nature of thought it is very desirable to ascertain the attributes that are proper to thought and are presented to our minds in every instance of thought. To inspect any such instance of thought, that is, to bring before our intuition the attributes that are presented when we think, and that may be discerned in every thought, to note the attributes thus presented, to group them into classes, is one of the leading necessities in a complete psychology.

To collect these attributes into classes and thus frame a system of the categories of thought, has been from the earliest days of philosophy a zealous labor of the ablest thinkers. We have as the results of these labors the Hindoo system of categories, the system of Aristotle, as well as modern systems, among which ranks most conspicuous that of Kant. These systems are certainly but approximations to an ideal perfection and have been, each in its turn, subjected to severe criticism. They have been condemned and reprobated especially, as was to be expected they would be, by men who had not carefully ascertained what a category as a class of attributes means. Mr. J. S. Mill, thus, in his work on Logic, strangely supposing that the categories were attempted "enunciations of existences," substitutes for the categories of Aristotle which he styles an "abortive classification of existences"—a system that confounds things, subjects, and attributes. His enumeration is, (1) Feelings; (2) Minds; (3) Bodies; and (4) Successions and Co-existences, Likenesses and Unlikenesses; a remarkable jumble of heterogeneous things as incomplete as confused, and altogether illogical and unsatisfactory. A system of categories of thought is simply a systematic collection of the attributes pertaining to thought. In forming it we are to proceed just as we would in forming a system of the attributes belonging to external bodies. In this latter case we take some particular body—an orange—and note the attributes presented to our perception and gather these into classes, as in Hamilton's enumeration of *extension, incompressibility, mobility, situation, attraction, repulsion, inertia; sense, impression.* Just in an analogous way we take a thought in its simplest form and note what attributes are presented to our intuition. The enumeration which we subjoin may not be complete; it would be presumptuous to claim such perfection of investigation in this stage of psychological science. But we may in our measure do a satisfactory work for ourselves if we proceed as far as our ability will allow, carefully and in scientific method.

§ 176. Reverting to our type-form of a primitive thought—*the orange is round*—we recall its origin in a perception in which an object, single and simple, was apprehended by the mind, being presented to it through the external sense. The perception was an incomplete stage of intelligence. The mind pressed on to a completed stage. The transition might be immediate, so that the thought should be simultaneous with the perception. It might, however, be prolonged more or less, or, indeed, possibly be broken off so as to be followed by no proper thought. The completed stage of intelligence gave us its discursive form in the judgment, the primitive form of thought; the essential characteristic of which, distinguishing it from the perception, was found to be that the single object in the perception was now discriminated into the twofold form of subject and attribute which, however, the mind still kept as one and identical—the subject not being a different reality from the attributes, but the same. The copula, being the identifying element, is thus essential to all thought and must characterize every valid derivative of thought. As the mind thus in reflection turns in thought upon what is given as one in the perception, it may recognize a second attribute or a third, in fact an indefinite number of attributes—intrinsic, as *yellow, juicy, decaying*, or extrinsic, as *present, selected from a number*, and the like, essential or accidental. Such are attributes pertaining to the object presented to the mind in the perception and originating in that. But there is another more important class of attributes which originate in the thought itself and pertain to thought as such and therefore may be recognized in any judgment whatever. These we now proceed to enumerate and unfold.

§ 177. 1. CATEGORY OF IDENTITY. In the first place, in reflecting on the judgment, *the orange is round*, we recognize the truth that as the copula, which identifies the subject and the attribute, is of the very essence of this judgment and of every judgment, and consequently of all thought or completed knowledge; everything of which we can think must admit this identification. There is given in this, as in every judgment, this attribute of *identity*, pertaining to whatever may be thought by the human mind, as every such object must, in order to be thought, present a subject that can be identified with an attribute or be differenced from it.

<small>Its origin.</small>

We have thus what is called the category of *identity*—the most fundamental of all the categories of pure thought. It will be remarked in regard to it, first, that it is an intuition. It is not a perception; it does not belong to the sensation; it does not belong to the external object—the orange. It is presented to us by the act of the intelligence in its completed form of a judgment. It is an intuition.

<small>An intuition.</small>

Next, it is a necessary attribute, in the sense that it is impossible to think at all without having this attribute pres-

<small>Necessary.</small>

ent in the thought, although seldom perhaps brought out into distinct consciousness. It is as necessary to thought as form and color to the orange, or to any visible object. It is in truth the essential characteristic of thought.

Further, it is universal, for no thought is ever experienced without it.

Still further, it is presented to the mind in a way precisely analogous to that in which the form and the color of the orange are presented. It is in the thought, and the mind takes notice of *How presented.* it; as the form is in the orange and the mind takes notice of it. We call this notion of the attribute in the one case an intuition, in the other a perception, simply to distinguish the different sources from which they come. The intuition comes from the thought within; the perception from the orange without. There is no mystery about the one more than about the other. The one is the attribute of an inner experience, the other of an outer object, both being founded in the nature of objects—the nature of mind and body. It is not to be assumed as if it had no ground. Thought itself is such that it has this attribute; as the orange is such that it has this round form. It misleads in regard to its nature to speak of identity as a native cognition of the mind, for it is no more so than figure in a visible object is a native cognition; or to speak of it as an original picture, independent in its rise in the mind of actual experience; or as a regulative law in any other sense than that every essential attribute of an object is a regulative law of that object. Identity is a regulative law of thought, simply because it is an essential attribute of thought—of thought as an actual experience, as a mental phenomenon.

Its opposite.
The opposite of identity is *difference.* The mind separates a subject from a supposed attribute as well as unites; it denies as well as affirms. These are the two alternative forms of thought—identifying and differencing; affirming and denying. We have accordingly two kinds of judgments, affirmative and negative.

Total or partial.
Identity is total or partial. It is total when the subject and the attribute which are identified in the judgment are in all respects one and the same, as I = I, or *the self is the self.* It is partial when the subject is only in part, in some respects but not in all, the same as the attribute. Most actual judgments are partially identical. In the judgment, the *orange is round,* the subject *orange* is identified only in respect to its form as *round*—is identified with but one of its manifold attributes.

Partial identity is denoted by the terms *likeness, similarity, resemblance.* These terms denote attributes, and as such ever imply subjects and are properly applicable only to subjects, whether original subjects or attributes, treated as subjects. We say: *An orange is like a peach in its form;* we do not say, *round is peach-like;* while we might say *the roundness is peach-like.* Similarity and resemblance as the etymology indicates, is but partial sameness or partial identity—sameness in some respects, not in all.

We have thus divers modifications of this attribute of identity. Besides those that have been mentioned there are manifold other intrinsic modifications, and of the extrinsic modifications, or those which are relative to the objects of thought or to other mental phenomena, there is obviously an indefinite number. All these modifications are grouped together under the one class and known by the name of the category of identity.

§ 178. 2. CATEGORY OF QUANTITY. It is obvious from an inspection of our typical judgment, *the orange is round*, that there is presented to us in a way perfectly analogous to that in which the attribute of *round* was presented to our perception and precisely as the attribute of identity was presented to our intuition, a second attribute of the judgment, which is designated under the name of *Quantity*. There is the one subject and there is the one attribute; these are different in a certain respect; at least so far as this, that one is subject of which something is thought, and the other is attribute which is thought of the subject; they are thus two, and yet they are one; they are identical in a certain respect; they are in fact identified in the very nature of the judgment. Two things—the two terms—are united. This property of being known as more than one in such a way that the several parts may be united in one whole, is the essential property of quantity.

Such is the rise of the category of Quantity. It

originates in a judgment as the completed stage of the intelligence by which an object given as single in the perception is recognized in the judgment as being in the two-fold form of subject and attribute, which terms, constituting the matter of the judgment, are identified in it. Perception gives no quantity; this is a property of thought. All thought is thus necessarily quantitative We may discern this attribute in every thought, every judgment, as we discern the attribute of *round* in the *orange*. It belongs to thought as thought; it characterizes all thought. It pertains, it should be observed, to the matter of thought—to the terms.

It is not an original, independent principle, existing by itself in the mind, or arising in the mind by any law of its nature, otherwise than as a simple attribute of thought. It is originally presented to us as an attribute of thought in the thought itself, precisely as the attribute of *round* is presented to us in the perception of the orange.

On this simple notion of quantity as one of the essential attributes of thought, to be recognized in any complete form of thought, rest all the modifications of quantity as diversely applied in the manifold processes of thought.

Whole and parts. But one form or derivation or application of this attribute should be specifically mentioned—it is the relation of whole and parts. The two terms of a judgment are parts which are in the identification of the judgment brought into one whole. This relation of whole and parts has thus its origin in the judgment. Whenever we think of an object as being a whole having parts, or as a part of a whole, we think such an object under the category of Quantity.

§ 179. 3. CATEGORY OF MODALITY.

Origin from the copula. A third attribute of thought is presented to us as we turn our view on the more essential element of thought—the copula—which identifies the two terms or parts of the matter—the subject and the attribute; it is the attribute of *modality*. The copula, or the proper thinking element in the judgment, is in this respect distinguished from the matter or the terms of the judgment. While it is our own, and is presented to our intuition by the mind itself, the matter of the judgment may be originally from a source foreign or extrinsic to the mind, and in every individual instance of thought is but a datum—something presented to the thought.

Now, as thought cannot deny itself, it must ever accept its own action. To thought itself the identifying element necessitates its own affirmations or negations. The matter as foreign to thought, is in reference to our thinking, accidental, contingent. To question whether the subject and the attribute are identified in valid thought is absurdity itself. The skeptic who questions this destroys the very foundation of all thought, of all opinion, of all belief, of all knowledge; and has no right to think, much less to question the thoughts of others.

Thus we have given in the very nature of a judgment the distinction of the necessary in thought from the contingent. But knowledge characterized in this respect as necessary or contingent, is thereby brought under the general category of *modality*; just as an orange characterized as round comes under the general attribute of form.

The leading forms in which this attribute appears are such as *possible* and *impossible ; probable* and *improbable ; necessary* and *contingent.*

As in the case of the other two categories, this one of modality is applied to objects external to the mind. Just so far as such objects approximate the nature of thought in this respect, they are regarded as necessary or the opposite. We speak thus of the necessities in nature as we speak of the necessities in truth or knowledge.

Modality, it is to be remarked, in its different modifications of necessity and contingency, is not an independent, self-existing thing; it is an attribute, and properly and originally an attribute of thought. It originates in that; it is a necessary property of all thought, as all thought, all true knowledge, ever admits of being regarded as necessary or otherwise.

§ 108. 4. CATEGORY OF PROPERTIES AND RELATIONS.—It is manifest from from the very nature of a judgment in which an attribute is identified with its subject that there must be in every object which can be thought that which will admit of being attributed. Every such object must have this character that it is *attributable.* We cannot think the orange to be round unless the orange, so far as we think in

<small>Origin.</small>

regard to it, so far as an object of thought, has this character of being attributable; of allowing in other words some kind of an attribution in relation to it.

All thought thus necessarily involves this general attribute of *attributableness*, and ever reveals or presents it to our intuition when properly turned towards it. But as we have found all attributes to be intrinsic or extrinsic, § 170, 171, and so distinguished into the two classes of intrinsic and extrinsic, otherwise known as properties and relations, to avoid the cumbrousness of the more exact and fitting name of attributableness, this category is called the category of properties and relations.

Designation.

Just so far as an object of thought is regarded in this view of having attributes, that is, properties or relations, it comes under this category.

We have distinguished properties into the two species of properties of quality and of action, § 172. The ground of this distinction we are now enabled to indicate. We have recognized under the category of quantity the category of whole and part.

Classes.

Intrinsic attributes are given us by the object itself as contained within it, and as subsisting in the object as a whole by itself irrespectively of all other objects. Extrinsic attributes are given us in the relations of the object as a part to the whole or to other parts.

But further, intrinsic attributes may be thought as wholes by themselves or as parts. We have thus

the two species of intrinsic attributes or properties:
—Qualities, which are intrinsic attributes regarded without reference to other objects, as *round, heavy;* and Attributes of Action, which are intrinsic attributes regarded in reference to other objects, as *rolling, gravitating.*

Attributes of quality are generally and normally expressed in grammatical adjectives; those of action are properly expressed in grammatical verbs combining the copula and in participles which by themselves do not combine the copula with the attribute.

§ 181. 5. CATEGORY OF SUBSTANCE AND CAUSE.—In the same way as we recognize in every instance of thought the category of attributableness embracing the two grand classes of attributes—properties and relations,—we also recognize, by turning our view to the subject or first term of the judgment, the general attribute, belonging to every object of thought, of its being a subject. The judgment *the orange is round* presents to our intuition this attribute of its having a subject to which something is attributed.

Origin.

Inasmuch now as any subject in thought may have an attribute either in the form of quality or of action, subjects are in this relation distinguished into two distinct classes: (1.) *Substances*, which imply that as subjects they take attributes of quality; and (2) *Causes*, which imply attributes of action.

Classes:—
1. Substances.
2. Causes.

This general category of substance and cause accordingly embraces the two subordinate categories, that of substance and that of cause. By some writers these categories, with perhaps more propriety but with a little more clumsiness, have been named *substantiality* and *causality.*

They have manifold modifications all embraced under the general category.

It should be remarked that the terms *substance* and *cause* are here used only as pertaining to thought. If we assume merely the fact of a judgment, no matter how it came to be, whether occasioned by the presence of some external object affecting our sensual organism, or by some inner condition of our bodies, or even by a direct touch of the creative spirit moving on our spiritual nature directly and through no sensuous medium,—if we assume simply the experience of a judgment, this category of substance and cause necessarily appears. As certainly as there is a subject in every judgment, just as certainly is there this attribute belonging to every object of thought, that it admits of being thought as a subject and either as substance or cause. The question of the existence of an outer world, of any thing truly actual or real except simply the existence of the judgment, is not involved at all in the admission that this category is given in every thought. In truth, as a category, that is, a collection of attributes, the idea of existences is excluded. Not therefore here by substance and cause as names of a category are we to understand any actual entity, any real existence.

Origin.
§ 181. 6. CATEGORIES OF THE TRUE, THE BEAUTIFUL, AND THE GOOD.—We

have, in this attempt to evolve the categories or classified attributes of thought, from the primitive form of thought, the judgment, assumed this judgment as a phenomenon, as an experience. We have accordingly assumed that the experience arose from some object being presented, which although single in itself when presented, the mind in thinking resolved at once into the twofold aspect of a subject and attribute. We have given us accordingly the two stages of mental experience which we have before recognized, the presentative and the reflective. Thought therefore necessarily supposes these stages ; and so far as the object is regarded as simply presented before being properly thought, the object is recognized still as one and simple and the mind apprehends it merely in its *form*. In other words, as has been before explained at length, it regards the object in so far as it is susceptible of being presented to a mind, as *beautiful*.

In the second or completed stage of knowledge in the judgment, moreover, the single object is resolved into the two elements of subject and attribute, which yet are identified in the judgment as one. The object thus regarded, no longer in its mere form, but in its essence as something that has attributes, is regarded as *true*.

Still further, the object has come to us ; has presented itself to us ; it has moved our sensibility ; it has prompted our thought. In this view of it, it is no longer a mere form, no longer a mere essence having congruous attributes, it is a power upon us ; and so far as it is regarded as thus moving us in harmony with our own susceptible and active nature, it is *good* to us. Looking thus at the effect of the object upon us we necessarily bring it under the category of *the good*.

It is thus apparent that every object which can become the matter of a judgment must in the different phases in which it is regarded, come respectively under each of these categories of the beautiful, the true, and the good.

CHAPTER VI.

EXISTENCES.

Origin of the notion.
§ 182. If we have given to us a thought—a judgment, as that *this orange is round*—we have found that there are presented to our intuition, if at least applied to it, by the very necessities of our thinking nature, by the necessities of thought itself as a completed act of intelligence, certain attributes, which we have in the preceding chapter gathered up into classes called the categories of thought. These attributes which thus necessarily belong to all thought, must belong to it if thought were nothing more than a mere supposition. But we are here to view thought as a fact. This fact is presented to our intuition as the orange to our perception. We know that we think, if we know anything. This fact of intuition, one and simple in itself, as we proceed to think of it, we recognize as a subject with attributes. These attributes, so far as essential, we have enumerated and grouped under the categories. But this thinking is itself an attribute. It is an action and necessarily implies a subject to which the attribute

belongs. We might construct a system of subjects of thought corresponding to these several attributes; that is, we might view these attributes as reflected to us from the light of their respective subjects, as in *shining* we recognize *something* that shines—*a Sun*. The fact of thinking would thus be distinguished into as many kinds of thought-subjects as there are of thought-attributes. But this would be useless abstraction. Language would fail us; for it does not fully distinguish the subject in a judgment from the substance or entity which in thinking we had resolved into subject and attribute. *Subject* and *substance* are, as we have found, used to some extent at least, interchangeably. It may be of importance, however, in unfolding the full nature of the intelligence in its completed stage, that of thought, to trace out under the general leading of the categories the classes of substances which constitute the objects of our thought.

We shall denominate these *existences*, meaning by the term the objective realities which are presented to our intelligence, so far as they can be generalized under the ascertained categories of thought. By the term is meant, then, not the subject in a judgment, but the object which the judging act resolves into a subject with attributes.

Entity.
The term *entity* is nearly or quite synonymous with Existence in this sense of an object.

Being.
The term *Being* has been used to denote the object of thought as a reality or existence. But this term has several differ-

ent meanings, the confusion of which in metaphysical discussions has been probably more prolific of serious error than has come from any other single source. In the first place, *being* is used to denote the mere relation between subject and attribute. In the judgment *the orange is round*, the word *is* denotes only the relation of agreement or identity between the subject *orange* and the attribute *round*. It imports of itself nothing of objective reality, yet nothing hardly is more common in metaphysical writings than the confounding of this mere subjective being, this mere being of thought, this being which is nothing more than the identification of the terms of a judgment, with actual existence. The reasonings of even so acute a thinker as Sir William Hamilton are not unfrequently vitiated by this fallacy. He over and over again deduces objective reality from the mere copula; as when, for instance, he makes the *cogito* of the famous enthymeme of Descartes: *cogito, ergo sum*, to mean *I exist thinking*. Whole systems of speculation rest on this fatal confusion of the *being* of thought with the *being* of objective reality.

In the next place the term *being*, like *existence*, is used both as a subject-word and as an attribute-word. Hence in abstract discussions arises a liability to confusion and consequent error.

Fichte assumes at the start an existence— the *ego*—; and assuming it as absolute, presents it as becoming known to itself by being limited or checked in its primitive absolute and so unconscious flow. That which thus checks its unconscious flow must be in some sense other than itself—must be a true *non-ego*. But this assumption of an absolute *ego* has no foundation and hence no warrant, except simply that it may help to explain the phenomena of thought. In assuming a judgment, on the contrary, we assume an unquestioned experience; and we have a solid and valid foundation for the entire superstructure of mental phenomena.

Assuming, accordingly, a judgment, we assume necessarily with this as essential to every judgment an attribute and with the attribute a subject, and with these, the object which is recognized in the judgment under the form of a subject with attribute. If the judgment be a reality, then the judging, the identifying of the subject with the attribute is a reality, for this is of the essence of the judgment. And this judging is an action, that is, as we have found, an attribute of a subject, which attribute and subject imply a substance. Something that judges is, therefore, given in the very fact of judging;—that is, an existence which is real, if the judgment be a reality; an existence too which is objective, for it is the object judged without which the judgment could not be. A real objective existence is thus given in the reality of any judgment. The famous enthymeme of Descartes: *cogito, ergo sum*, taken impersonally as it should be in this sense—there is thinking, therefore there is existence—is irrefutably sound and valid. The premise is a reality—there is thinking. This is assumed. He who denies the premise, that is, he who denies that thinking is real, and he only, can avoid the conclusion that there is real existence. He who maintains that thought is a mere phantom, or that it is a mere flux with nothing that flows, may consistently deny the conclusion that the fact of thinking, proves a thinker, and, at the same time, also an object thought. But let it be granted that there is real thinking and the proof of objective existence is beyond question.

Further, the fact of thinking proves a plurality of existences. For, as we have seen, if there is thinking, there is both an existence that thinks, and also an existence that is thought. For the mind cannot make itself an object to itself, until it reveals itself by some action ; and such action, as we have before seen, must in the case of a finite mind, be occasioned by some object external to itself. To this the category of Quantity at once leads us ; and in the expanded form of whole and part, giving the category of attributes of Relation, it conducts us necessarily to an indefinite number of existences, inasmuch as each part, which can be one term in a relation, must be recognized as one object of itself.

Of this indefinite plurality of existences we now proceed to enumerate the more fundamentally generic classes.

§ 183. First, as already shown, we have **Reality of mind.** given us in the fact of thought a thinking existence—*a mind* or *spirit*.

§ 184. Secondly, as already seen, we **Of motion.** have also given us in the very fact of thought an object external to the thinking mind—a *not-self* as contrasted with the mind—*the self*. Of the distinctive nature of this existence external to mind, this *not-self*, thought in itself is incompetent to testify farther than this, that it must be capable of communicating itself to *the self*. If, however, we admit together with the thought, the fact of a sensibility as the medium through which external things introduce themselves to our minds, then we necessarily conceive of all such communicating objects as having form or body in its largest import, embracing immaterial as well as material body. If, further, we admit a sensuous or bodily organism as the medium through which our minds apprehend external objects, we come to the class of existences termed *material*. In this way we reach the truth that there is matter distinct from mind, and that this matter serves as a body to idea for inter-communication between minds.

§ 185. Thirdly, inasmuch as all objects presented to any one mind must be so related to one another as to be thought by a single mind, and so capable of being aggregated together, we come, under the category of whole and parts or that of relation, to the notion of an existence, that is a whole, embracing as its parts all the objects of which we think—a proper *universe*.

The universe.

§ 186. Fourthly, placing ourselves under the lead of the category of substance and cause, and admitting the reality of the objects of thought, we are conducted directly to those classes of existences called respectively substance and cause — a substance

Substance and cause.

being an existence thought under the attribute of Quality; a cause being an existence thought under the attribute of Action. Every existence is both a substance and a cause ; but as we think any existence under the one or the other of these categories it becomes to our thought a substance or a cause. The sun thought as *heavy*, is a substance ; thought as *attracting*, is a cause. An orange is a substance when regarded as *juicy;* it is a cause, when regarded as *refreshing.*

§ 177. Fifthly, under the category of Condition we think an attribute in reference to the whole without reference to its having parts ; while under the category of Relation we think an attribute to the whole as having parts. As this latter category has conducted us to the collective existence which we call the universe, in which term we have distinct reference to the parts which make up the whole, so the former category, that of Condition, directly conducts us to a whole without reference to parts, that is objects in it and constituting it. Such a whole may be thought by us either under the category of Quality when we have the idea of *Space*, or under the category of Action when we have the idea of *Time*.

[Marginal note: Space and time.]

In regard to these ideas of Space and Time, their nature and origin, it would be presumptuous to dogmatize in the present stage of metaphysical science. Philosophers of the highest rank insist that they are mere forms of human thought—schemes of the human understanding ; while others of equal rank, as do the mass of men, hold them to be real existences. We may safely advance the following propositions in regard to their nature and origin:

1. The ideas of Space and Time are not objects of perception. They are given only to our intuition in the fact of thought.

2. The mind is naturally pointed to them under the category of condition, which we have found to be a necessary attribute of thought.

3. The category of condition is immediately grounded on the objects of thought—on the terms or matter of a judgment—not on the copula element. They have therefore not a pure subjective, but a proper objective nature.

4. Admitting that thought is real and its objects real, we cannot resist the conclusion that the category of condition partakes of that reality—is a class of real attributes, and consequently that the class of subjects or substances corresponding to this category of attributes is, in some true and proper sense, also real.

5. Space is not mere extension, for extension is an attribute of a part; whereas space implies in its very nature a whole. In the same way, Time is not mere succession, for this respects parts; whereas Space and Time are given us in thought as under the category of condition, which regards wholes irrespectively of their parts.

6. Space and time are shown in their very origin to be unlimited. As presented to our mind under the category of condition, they are as unlimited as the universe of thought. All those reasonings, therefore, which imply limitations in the nature of

space or time lead at once to contradiction and absurdity.

Human thought as the function of a finite mind is indeed limited; it cannot grasp the universe. But this imports no limitation to the universe itself; for even to finite thought there can be no boundary, no limit assigned in the universe of objects from beyond which no object can be presented. Just so, space and time are unlimited, although the ideas of space and time originate in finite thought, because no line can be drawn beyond which we can think no space; none beyond which either in the past or in the future we can think no time.

CHAPTER VII.

INTELLECTUAL APPREHENSION AND REPRESENTATION.

Twofold modification of intelligence in respect to sensibility. § 188. The intelligence is modified in a two fold way in its relations to the sensibility, (1) as receiving truth, and (2) as expressing truth. The intelligence is thus both a capacity of knowledge and also a faculty of knowledge, just as the sensibility is passive as recipient of form, and also active as productive of form. In its function as a capacity, the action of the intelligence is denominated *Apprehension;* in its function as a faculty, it is denominated *Representation*.

1. Apprehension. § 189. INTELLECTUAL APPREHENSION is accordingly defined as the function of the mind in receiving the true.

It includes Perception, Intuition, Thought or Conception, in so far as they are modes of receiving or acquiring knowledge. They are the different kinds of apprehension distinguished in reference to the two sources of knowledge, external and internal, and the two stages of knowledge, introductory and complete.

It is distinguished as *comprehensive* when the

grounds of the truth are also apprehended in connection with the truth itself.

It is related to the sensibility on its passive side. It is in truth, as before intimated, the intelligence side of the single mental state which, when regarded on the side of the sensibility, is named under the forms of that function. Perception, thus, is but the intelligence side of sensation.

This mental act or state, is modified indefinitely in respect to the relative degrees in which the sensibility and the intelligence appear in it. The one or the other may greatly predominate in different cases. Yet even when the sensibility is predominant, we may direct our attention rather to the intellectual side, and so make it the really predominant element to our view, and then we speak of the act as apprehension and not as impression or affection, or other term denoting properly a form of the sensibility.

2. Representation. § 190. INTELLECTUAL REPRESENTATION is defined as the function of the mind in presenting the true.

It is distinguished as *Demonstration*, in the larger sense, when the grounds of the truth are presented with the truth itself.

It is related to the active side of the sensibility, the imagination proper. It differs from Philosophical Imagination only in this, that it points to the intelligence side of the act or state, while the latter term points to the imagination side. § 143.

The two functions vary indefinitely in their relative degrees of predominance in different cases.

CHAPTER VIII.

CURIOSITY AND ATTENTION.

Twofold modification of intelligence in respect to will.
§ 191. The Intelligence is modified in respect to the will in a twofold way: (1) as determined in its activity only by the instincts of mind as an essentially active nature, in *curiosity;* and (2) as positively determined by the will proper, in *attention*.

1. Curiosity.
§ 192. By CURIOSITY is meant, the instinctive desire of knowledge in the human mind.

The mind, so far as active, seeks for truth or knowledge. This feature particularly characterizes it in its infancy. All objects of knowledge are alike attractive to it, for its selecting power is then feeble, and habit, taste, or disposition is undeveloped. In its progress, this instinct, unless overborne in indolence or indulgence, acquires ever additional strength. If rightly directed and cultivated, it ultimately makes the intellectual giant in knowledge. With advancing development, it turns more and more to specific fields of truth and acquires distinction in particular branches of knowledge.

CURIOSITY AND ATTENTION. 211

2. Attention.
§ 193. By ATTENTION is meant the voluntary determination of the intelligence to objects of knowledge.

Curiosity passes into attention in the natural growth of mind as instinct passes into power of will: and the mind acquires in its growth more and more entire and absolute control over its own acts and states. The desire of knowledge—curiosity—at the same time strengthens in itself, and also " spends itself in will." Attention is susceptible of indefinite development. It is very weak in beginning study. The tyro in knowledge finds it hard to keep his thought steadily on any subject of study. The power of attention grows as he advances. In its higher degrees it marks the intellectual genius; for nothing more characterizes the man of genius than the power of fixed attention.

Attention is conscious or unconscious. At first, it is necessary that the mind with deliberate, conscious intention, bend itself to its work, exclude distracting objects, and fasten its regard on the single subject of its study. Repeated effort in this conscious attention passes into habit; and the mind holds on in its attentive study, conscious of no particular energy of the will.

Attention, as applied to external objects, is known as *Observation;* as applied to matters of our own consciousness, it is designated *Reflection.*

Observation, it should be remarked, includes both stages of cognition—perception and judgment; as also reflection includes both intuition and judgment.

BOOK IV.

THE WILL.

CHAPTER I.

ITS NATURE AND MODIFICATIONS.

Will defined. § 194. THE WILL *is the mind's faculty of choosing*.

Synonyms. This faculty is otherwise known as the Voluntary Power, the Orectic Faculty, the Conative Power, the Moral Power, the Power of Choice, the Free Will, the Faculty of Freedom.

Its product is called a choice; also, a volition.

Fourfold modification. It is variously modified, as first, in relation to its object. As essentially active it ever regards an object in its action. This object is known under the name of *Motive*.

Secondly, in relation to its stages, as (1) incomplete and prelusive, and (2) as complete.

Thirdly, in relation to its own growth and its special subordinations.

Fourthly, in relation to the other mental functions.

We will proceed in the order of this enumeration to speak of these several classes of modifications, after having first presented the intrinsic attributes of will as given in experience.

CHAPTER II.

CHOICE.

Choice exemplified. § 195. It will not be difficult to identify an act of will—a choice, a volition—and to distinguish it from other acts or states of mind. When an orange is presented to me and it impresses my sense, I am only passively affected; I have a feeling. When I recognize it as having certain attributes, so that I can say *it is round, it is sweet,* my intellectual activity is engaged; I have a knowledge—a cognition, of the object. But I may, besides feeling and knowing the object, put forth an effort in reference to it; as, for instance, to take it. I now exert my will; I have a choice, a volition. The feeling and the knowledge have preceded my willing to take it; I may choose the orange in full intelligence of it and of what I am doing and also in expectant feeling or desire; but the act of taking is beyond, and, in our thought at least, distinguishable both from the feeling and the knowledge. The act of taking is somewhat complex, embracing something more than a mere choice. But we cannot fail to recognize in this complex act something more than a feeling or a knowledge ; and the characteristic element in the act is a choice or volition.

The act of taking embraces in fact two very distinguishable elements. There is, first, the determination, the decision, the purpose, to take the orange ; and there is, besides, the putting forth of the hand to take it. The first of these elements may be a completed act of will, a full choice or volition, as the blossom may be complete, although followed by no fruit. It might exist even although it were followed by no reaching forth of the hand ; although at the instant of determining to take the orange, all the bodily functions were paralyzed. This second act, putting forth the hand, is simply executive of the previous choice; it is called, in distinction, an *executive volition*.

§ 196. If now we carefully study this determination of the will this choice, we at once recognize it as *an act*. It is the exertion of a power, of an active nature. In this it differs essentially from a feeling, a mere passive affection of the sensibility. All choice is essentially *active*.

<small>An act.</small>

§ 197. Further, there is clearly an appropriating, a making of our own, at least in intent, of the object.

<small>Appropriative.</small>

Herein, a choice differs essentially from an act of the intelligence which possesses no such characteristic. In knowing, we only possess subjectively ; we have, we possess, a thought, a knowledge ; we do not possess the object itself. But in a choice we possess, in intent at least, the object itself. An act of will, thus, is ever appropriative. It ever seeks to appropriate its object and make it its own.

Free.

§ 198. If now we take an instance of a peculiar kind of choice, in fact, a true choice or act of will, full and entire ;— if, for instance, we suppose the orange, which we have determined to take, not to be our own, but another's, who refuses us his permission to take it, and if we still determine to take it by stealth or by violence, we discover in our choice another class of elements. We discover, first, that such a choice involves *freedom*.

It is implied in this that the determination to take was not forced upon us by any insuperable necessity; that it could be withheld as truly as be put forth. We never think of saying, however pressed, in self-vindication, that literally we could not help taking it ;—that we were necessitated to take it. We are conscious that in every such act we could take or forbear taking ; we could choose or refuse. Accordingly we acknowledge our responsibility for the act. To deny this element of freedom in many at least of our choices, is to belie the testimony of our own consciousness ; it is to contradict the universal testimony of intelligent and unbiased men ; it is to falsify the universal language of man, which in all its dialects comprises terms importing this freedom in choice. An act of will, choice, volition, in its highest and most proper form at least, involves then the element of freedom.

§ 199. Another of the higher elements involved in proper choice is distinct *personality*.

Personal

This element is indeed dimly given in feeling and in knowing. The phenomenon of feeling gives the distinction of an object impressing and a subject impressed; as does that of knowing give the distinction of object known and subject knowing. But this elementary and germinant distinction of personality rises into perfect outline and fulness in the free will, with an emphasis not allowable before. The feeling and knowing subject in willing, recognizes and pronounces itself a true *ego*, a person distinct from other persons and things.

But this free personality which has its seat in the will and constitutes the leading and characteristic element of that mental power, itself involves several distinguishable attributes of highest interest and importance.

§ 200. First, free personality involves *mental sovereignty*.

Sovereign.

The free will rules over the whole soul, holding the sensibility and the intelligence in strict subjection to itself and under its own control.

This mental sovereignty residing in the personal free will of man is by no means absolute. The very finiteness of his being, which we have so fully recognized, forbids this idea. The domain of the will is limited both outwardly and inwardly. It is limited in relation to the universe of beings and objects and activities without itself. Its utmost exertions soon reach a limit beyond which they cannot be pushed. It meets even within its own proper limited domain with checks and obstacles which it often finds itself unable to overbear or remove. Man's universal experience leaves recorded in the consciousness the clear, salient characters of the dependence and finiteness of the human will.

This sovereignty of the human will is limited, also, in relation to the mind itself of which it is the chief function. Its power does not reach so far as to reconstruct the mind or change its essential attributes. It cannot make the sensibility feel, the imagination form, the intelligence know or apprehend or represent, otherwise than according to their own nature and laws. It cannot utterly destroy, if it may impair, the essential activity of the soul. It cannot prevent its feeling or its knowing. It cannot abrogate utterly its own freedom, or its own activity, however much it may weaken, corrupt, or hamper its proper function and character.

But while thus dependent and limited in its sovereignty, the personal free will is a true sovereign. It rules the sensibility. While it cannot prevent feeling when an object is presented to the sensibility, and cannot remove from the reach of all objects that can impress it, inasmuch as it cannot remove itself from the universe of being,—cannot altogether prevent feeling—it can yet direct feeling in various ways. It can, subject indeed to the power above it on which it depends, select the objects to which it will allow access to its sensibility. It can arrest feeling when going out towards any one object, and turn it towards another object. The angry man expels his wrath by bringing before his sensibility an object of fear or of love; by closing his eye on the provocation to anger and opening it on what excites compassion or gratitude or reverence.

In a much higher sense does the free will rule the imagination or faculty of form. It prescribes the idea to be formed, as well as the matter in which to form it, and prompts and directs and sustains the forming act. It is indeed the soul of the imagination as an active nature.

It bears a similar relation to the intelligence. It puts it in action; its elects the object; it arrests or sustains the activity.

The personal free will is thus sovereign in a true sense over the sensibility and the intelligence. It is equally sovereign, as will be shown farther on, over its own subordinate movements and the so-called executive volitions.

When the will acts in conjunction with the sensibility, the imagination, and the intelligence, in determining the objects of mental activity,—as in selecting objects of sense or forms of imagination, or kinds of attributes to be recognized, and especially in determining upon the ends or aims towards which the mental activity is to be directed—we discover a form of acting which has been ascribed to the so-called *regulative faculty* of the mind. It is, however, only a special modification of the function of willing by reason of the will's acting here through the imagination and in the light and thus under the lead of the intelligence. Man, as rational, has an aim in his action; the intelligent selection of this aim and the direction of the mind's action in feeling, thought, or endeavor in reference to it, is the special characteristic of rationality. Just so far as fully and completely rational, therefore, the human soul is regulative. It has, accordingly, no special faculty endowed with this regulative function, to be classed co-ordinately with the imagination and the intelligence.

§ 201. Secondly, the free personality involves the attribute of *originativeness*.

Originative.

In a sense in which it cannot be said of the sensibility and the intelligence, the will is a true originator. As part of a finite being, it is dependent on something external to itself for the object towards which its activity is to be directed. Free choice is in this low sense determined by its object as presented to it. There can be no choice where there is nothing to be chosen, as a man however strong cannot lift a weight unless there be a weight to be lifted. In a sense analogous to that in which we say the weight determines the lifting, we may say perhaps that the object chosen determines the choice. But there is a true sense in which the free will may be said to originate action. As the man determines whether he will lift the weight presented to him, so the free will ever determines its action in this or that direction to be or not to be. Freedom supposes ever an alternative of choice. If there be but one object presented there is the simple alternative of choice or refusal. If two or more objects are presented only one of which can be taken, the alternative is conplicated; the choice or refusal is combined with the act of electing or selecting the one or the other of the objects. We have in this case elective choice or refusal. Of the choice or refusal, whichever it be, and whether simple or elective, the free will is justly called the originator.

The free will of man accordingly is so constituted by its creator as to be able to enter the realm of mere nature as it flows on in its necessary flow and to originate new sequences beyond the power of mere nature. It does not originate new matter; but it does originate new dispositions of matter. It does not originate new measures of force; but it does originate new directions of force, so that the sequences of nature are more or less changed from their undisturbed order. It does not originate in the sense of exerting new choices or purposes in other free beings; but it does present to them new objects, new motives, new inspirations which may induce new purposes and character in them while still remaining in unchecked freedom.

Moral.
§ 202. Thirdly, the free personality involves the attribute of *morality*.

By morality is expressed the relation of a being to right and duty. By virtue of its freedom, of its freedom however as necessarily intelligent and feeling, the mind of man has rights which it exacts and duties which it owes. The mere animal has no proper rights, owes no proper duties. Right and duty are reciprocal; what is my right, is another's duty; what is my duty is another's right. The personal free will is the seat and centre of this relation of man to right and duty; and is the source out of which it naturally and necessarily springs.

§ 203. Fourthly, the free personality involves the attribute of *responsibility*.

<small>Responsible.</small>

The finiteness of man's being and his dependence already in themselves foreshadow a power above him, by which he is limited and hemmed in, and on which he depends. But in his free activity this relation to a higher power shines out clearly and in definite outlines. As the exactor of rights and the subject of duties, he recognizes a law from without and from above which has allowed those exacted rights and has prescribed those owed duties. He recognizes a law written on the very centre of his being, his inmost personality, that at once imposes duties and gives rights. He recognizes also a power to sustain and to enforce this law, to which he expects all other beings from whom he has rights to be answerable, to which acccordingly he feels they must expect him to be answerable, so far as he is bound in duty to them. The free personality thus makes man moral, as subject to a law which enforces duty and sustains rights.

It is important to remark that this characteristic of free personality, involves at once the distinction of the personal moral self from other personal moral beings. It involves, also, the recognition of a personal free being who is the source of the law of duty and equally its administrator. The responsibility of a free person must be not to a thing, not to an attribute, but to a free person. This free person we call God, who writes the law of duty on the human soul and rules to sustain that law. The free personality is thus shown to be the seat and centre and source of *religion,* the base of the relation of man to God

CHAPTER III.

MOTIVE

Motive defined. § 204. By a motive is to be understood the object of the will in its action. In other words, it is that in respect to which the will acts.

As essentially an active nature, the will must have an object in respect to which it is to act. This is that necessary incident of a finite and dependent being which we have all along been careful to recognize. This object must primitively of necessity be presented to it from without the mind. The action of the mind once awakened, however, may afterwards, revealing itself as it were to itself, present objects to itself in its function of willing Motives are thus presented in intuitions as well as perceptions.

In a certain low sense the motive may, as has been already stated, be said to determine the will. It is the object without which the action could not take place ; it also determines the direction in which the activity of the will goes forth. When I choose an orange, it so far determines the action of my will as that except for its presence my will would not choose it and also that my will moves towards it and not towards any other object around me; it so far determines selection and choice. In this limited sense the motive may be truly said to determine the will; it determines the possibility of action since every volition must have an object, and also determines the particular direction in which the specific volition moves. In the sense that the motive is the immediate source of the volition, or that it so affects the will that it has no freedom in the case, or that it and not the mind puts forth the volition, it cannot truly be said that the motive determines the will.

§ 205. An object can be such to the will only in so far as it is good. In other words, a motive must be a good.

A good.

Nothing but a good can be object to the will. The true, the beautiful, the good, we have seen, are three comprehensive ideas which include all possible objects to the mind. But the true we have found to be exclusively object for the intelligence, and the beautiful for the sensibility. There remains consequently nothing but the good as object for the will.

That good must constitute the essential nature of a motive is to be presumed from the goodness of the creator. In the will is centered the entire free personal activity of the soul of man. That the legitimate exercise of this free activity should lead only to good follows necessarily from the assumed perfection of God.

Experience and observation confirm this *a priori* conclusion. Even in willing contrary to the good law of our being, and so choosing evil, it is not the evil but some apprehended good in the evil that is the immediate and proper object of our choice. The very impersonation of evil, Milton makes to recognize the truth that good in some sense must be the object in all free action, as he makes him utter: "Evil be then my good." A divided empire with Heaven was the good he proposed to himself and chose. Even the purest malice thus must propose some good to be attained with all the evil which it may intelligently or blindly bring to itself by its resolve. There is a certain pleasure, a good, in revenge and even in unprovoked cruelty.

§ 206. A motive is a good to the mind, to the whole soul, not properly and strictly to the will.

To the whole mind.

The will is not properly the function of the mind by which it receives or experiences good. The good which makes an object a motive is for a capacity rather than a faculty; and the will is essentially a faculty. Much less does the will perceive the good in a motive. This perception belongs to another function—the intelligence. In so far as the mind feels good or perceives good in a motive, it is by its functions of feeling and knowing, not by its function of willing. All that language which represents the will as passively affected by the motive, or as viewing the good in the motive, must be taken figuratively as intended only to mark the will as in the place of the whole soul, or must lead to confusion and error. The will neither feels nor views. The true representation would be that when the mind by its function of feeling—the sensibility—feels the good in an object of volition, and by its function of perceiving—the intelligence—perceives or knows it, it may by its function of willing choose that object which as such felt and perceived good has become a motive to it.

In the mind. § 207. Farther, a motive, in so far as an object to the will, must be in the mind.

Loosely speaking we may sometimes speak of the external object in itself as the motive; but in strict truth a motive must ever be internal. The object must be a good feelingly and consciously apprehended by the mind before it can become a motive. In this sense is to be understood the doctrine of ethical writers, that a motive includes the object, the intellectual apprehension of the object, and the desire or affection awakened by it.

Classes. § 208. Motives are conveniently distributed, in respect to their original source, into two classes—*external* and *internal*.

1. External. § 209. AN EXTERNAL MOTIVE is an object of volition, originating from without the mind.

Thus an orange apprehended as good, and so presented to the mind as object of choice, may be called an external motive, because the immediate motive which is the orange as perceived and desired; or rather which is the desire of the orange as felt and perceived, originates in an object external to the mind.

2 Internal. § 210. AN INTERNAL MOTIVE is an object of volition originating in the mind itself.

The culture of one's own faculties presented as an object of choice is thus an internal motive.

Not infrequently is the will addressed at the same time by a motive of each class, and the alternative of choice lies between the two as opposed to each other. Sometimes the motive is complicated of both objects united in one, to be chosen or rejected together. Sometimes the alternative of choice lies between two or more motives of the same class.

In all cases, however, it is to be remembered that a motive as the object willed by the mind, lies wholly in the mind itself, whatever may be the fact in regard to its origin and history. It is ever a good as felt and perceived. The feeling may be an illusion, as when one suddenly awakened from sleep mistakes a shadow for a substance and moves to avoid or to assail it. The perception or the intuition or the full thought which presents the motive may be unreal, or more or less incongruous and incorrect; it is not the proper function of the will to prove the reality or the truthfulness of its motives. This lies in the sphere of the intelligence. It is, to restate the important fact, it is the feeling and knowing mind that by its functions of willing chooses; and that in choosing cannot but act with some measure of intelligence and of feeling.

CHAPTER IV.

GROWTH AND SUBORDINATIONS OF WILL.

Will a ruling faculty.
§ 210. In the mind as an essentially active nature, the will appears as the ruling and directive activity. It rules the other functions of feeling and knowing, and also, as we shall see, in all subordinate volitions it rules itself. It rules the sensibility by selecting the objects which shall be allowed to address it; by arresting the address of such objects after being once allowed; by yielding the sensibility more or less to the influence of its objects. In the same way it rules the intelligence by choosing its objects, and by directing and intensifying the attention. It cannot wholly prevent the mind from feeling or from knowing, for the nature of the soul cannot be annihilated in any essential element except by the power that created it. But by the choice and change of objects, and by allowing the mental activity to be more or less engaged, it rules both feeling and knowledge.

Growth of will.
§ 211. As itself participating in an active nature, the human will is susceptible of growth.

In infancy the will is feeble, bordering on impotency. By exercise it becomes mighty through the principle of habit and growth. It is developed out of the instinctive nature of the mind. The transition from action which is merely instinctive, and as such necessitated by the will of the creator in creating it, is beyond the notice of our limited observation. We can as well observe the development of the bud from its germ. But by the very law of all mental life its action once prompted continues on, and although in a sense changed in its direction or opposed by subsequent volitions, yet never can be truly said to lose its record in the mind's history. Each volition not only strengthens the willing mind itself, as legitimate exercise strengthens all living power, but each repetition of the volition in the same direction or towards the same object confirms the tendency to will in that direction. The will thus may acquire in time what in popular phrase we term indomitable determination; it is proof against all motive that finite power can bring to it. Weakness of will, in other words, imbecility of purpose, vacillation, irresoluteness, is the result of varying volitions, one moving in one direction, another in another. Strength of will, on the other hand, under the great law of growth comes directly and surely by multiplying volitions in the same direction, that is, towards the same or similar objects, and by shunning volitions looking in opposite directions.

Dependent. § 212. The will, however, as one function in a mind that is itself single, is so far dependent on the other functions of the sensibility and the intelligence.

The objects of volition as motives, without which the will cannot act, come to it through these other functions. And further than this, its strength is also dependent on them. A feeble sense, a feeble understanding, is attended by a feeble volition. In the intensest feeling and the clearest knowledge, springs ever the most energetic will.

The will grows thus by exercise, especially by exercise in the same direction, and as attended by a lively sense and a clear and rich intelligence.

§ 213. The will, as has been already stated, rules itself, in a certain sense, as well as the feeling and other functions of the mind.

Self-ruling.

It does this by putting forth volitions which draw along, whether more positively by its own free prompting and sustaining, or more negatively by allowing and suffering other following volitions. Such originating volitions are called *governing*, or *ruling*, or *predominant* volitions. The volitions which they respectively draw along after them, are called, in reference to the former, *subordinate* volitions. We determine, thus, for a single illustration, to take a journey. This determination of will is, in reference to the particular acts by which it is carried into execution, a governing or predominant volition. Every particular act of will put forth to carry out this original determination, of getting ready the baggage, procuring the conveyance, etc., is a subordinate volition. Such subordinate volitions, in so far as they are regarded as carrying out the governing volition, are called *executive* volitions. The putting forth the hand to take the orange after the determination to appropriate it, is an executive volition.

It is obvious that the same volition may be in one relation a predominant volition, and in another relation a subordinate volition. The getting ready one's baggage is subordinate and executive in relation to the predominant volition to take a journey; it is itself predominant in relation to each specific volition, as going to the shop to purchase, purchasing, ordering or bearing home, parting, etc.

The highest volition of which man is capable, and thus with him absolutely the predominant volition which is subordinate to no other, is that which controls the entire activity of the mind so far as subject to the will itself. Such a predominant volition determines the character of the man in its largest and most proper sense. From the very nature of motive or object to the will, such predominant purpose must have for its object as motive the chief good of the soul as actually selected by it. The good so taken to be the chief good may possibly be an inferior good, as compared with some other good. Such is the prerogative of the will as essentially free; it can choose the lesser of two goods. In the grand alternative of choice in which God is proposed as one of the objects and rejected or declined, the lower good is in fact chosen as the chief good. And this choice of the inferior good is the sin; as St. Augustine in his confessions B, ii., § x. well defines:—"Sin is committed while through an immoderate inclination towards those goods of the lowest order, the better and higher are forsaken." Such sinful choice, although of a lower good, yet becomes the predominant volition, and so governs and determines the following acts of the moral life, and characterizes the entire current of the soul's **activity.**

CHAPTER V.

CONSCIENCE.

Conscience explained.

§ 214. The term *conscience*, originally and etymologically synonymous with *consciousness*, denoted generally self-knowledge. But usage has greatly modified its signification, first by restricting it to matters of will or morality, and secondly, by enlarging it to include feeling as well as knowledge. It has, therefore, acquired a new import widely differing from its primary sense.

Other expressions are in use to denote the same mental state with more or less modifications, as *moral sense, moral faculty, sense of duty,* or *of right and wrong.*

Threefold elements.

§ 215. Conscience includes three chief distinguishable elements :—(1) a discernment of right and wrong ; (2) a feeling of obligation ; and (3) an approval or disapproval.

The term is used sometimes with more prominent reference to one of these elements, sometimes with more prominent reference to another. It properly implies, however, all three, even when used with such prominent reference to one, inasmuch as the three necessarily exist and imply one another.

§ 216.   1. Conscience involves, as a chief element, the discernment of right and wrong.

Discerning of right and wrong.

The rise of this complex phenomenon of mind is immediately and necessarily out of a conscious act of free will. We have already recognized the truth that the idea of free personality involves the idea of being a subject of rights and duties, that is, the idea of morality. In other words, the fact of free choice reveals to us at once the attributes of morality as truly as the orange reveals to us the attribute of form or of color. It is impossible for us fully to contemplate such an act without recognizing this attribute of morality, by which is understood that the act must be considered as either right or wrong.

This is the proper origin of the category of morality which includes under it the specific and alternative attributes of right and wrong. It is true, however, that the existence of this attribute as pertaining to free action may be proved from other assumed truths. From the assumption of the being and rule of God there follows by necessary deduction the subjection of his free creatures to him, which subjection implies then enforcement upon them of the observance of the right and the avoidance of the wrong. He cannot rule without subjects; and as he is free and righteous, he cannot be true to himself but as requiring right of his free subjects.

The existence of this attribute as pertaining to

free action may be deduced equally from the assumed existence of that true law, right reason or rule, invariable, eternal, universal, of which Cicero so profoundly and so justly discourses. Given such a law, and it follows that action under it must be characterized as right or wrong.

It may be proved also from universal acknowledgment, from the general consciousness of men, and especially as expressed in the language of men.

This attribute of free action—that it is moral, that is, either right or wrong—as necessarily pertaining to it, may be discerned by the human intelligence in every case, whether the act be one's own, and so properly within the range of personal consciousness, or another's and apprehended by observation.

The fundamental element in conscience is this discernment of the right or wrong in every free act which of itself and immediately reveals this attribute to every free contemplation.

§ 217. 2. In a similar way arises the sentiment of obligation.

2. Feeling of obligation.

A sense of moral freedom involves a sense of obligation to do the right and shun the wrong. So soon as a free choice is proposed, obligation is felt. As every free volition involves the necessity of an alternative determination, of choosing or refusing, or of selecting the one or the other of two objects, and as there is given in this freedom the attribute of being obligatory—of constraining to the right—so the sensibility is impressible by the attribute. It is true, the mind in its sovereign freedom, may turn away to a certain degree its sensibility from the attribute: yet as the mind is in its highest nature a free and consequently a moral agent, this sense of obligation cannot be utterly prevented or annihilated.

This sense of obligation, thus necessarily springing from the consciousness of freedom and of choice, has for its objective counterpart what is fitly called "the law of God written on the heart." It is accordingly a legitimate inference from this consciousness, from this sense of obligation, that there is an outer source of this obligation; that there is a law, given to man from without himself, and inscribed on his inmost nature; and that this source is none other than God himself, who created man and endowed him with his freedom and who wrote the law in his inmost being and rules ever to sustain and enforce it.

§ 218. 3. Still further, the full contemplation of an act of free-will necessarily brings along with it a sense of approval or of disapproval.

3. Approving or disapproving.

Every such act reveals in itself this attribute of awakening this feeling, as the orange reveals the attribute of juiciness and so impresses the outward sense. Relatively to the doer, and as seated in him, the attribute is that of merit or demerit, desert or guilt. In every free act the doer feels this desert or ill-desert according as he has chosen right or wrong, and exactly correspondent to this feeling in the heart of the personal doer is the judgment of approval or condemnation, of praise or of blame, by whoever scans the act with a moral eye.

Such is the three-fold function of conscience: it discerns in every free act the right or the wrong; it feels the obligation to do the right and to shun the wrong; it approves or condemns— awards praise or blame.

Conscience, it should be added, has sometimes been regarded as the seat of that pleasure or pain which attends on all mental activity, and which in moral acts and states is deepest and most intense. We speak of the pleasure of a good conscience, and this pleasure may, not unwarrantably, be regarded as a function of conscience. In this case we should add as its fourth function that of giving the sense of that peculiar pleasure or of pain in the doer which naturally attends all right or wrong action.

§ 219. The will extends its sovereignty over the conscience as over the entire mental activity.

Subject to the will.

It directs and controls the culture of conscience, which like all other mental activities is capable of culture and growth. Quickness and accuracy of moral discernment, tender sense of obligation, and ready and just response of praise or blame, are matters of culture. There is open to man a path of advancement, of ascent, leading ever on and up towards that infinite perfection which belongs to the judge and ruler of all.

The will, also, regulates and controls the conscience in respect to specific acts. Most moral acts of men are more or less complex, embracing some lawful elements, some unlawful. Morality in this respect is like truth and beauty; it appears among men like them in forms complicated of the perfect and the imperfect. As there is some deformity in almost every beautiful form on earth, some error in almost every truth held by men, so there is in every right act of man some taint of imperfection. And, on the contrary, there is no form wholly destitute of every beauty, no error void of all truth, no sin destitute of some feature that is morally approvable. The will can thus fasten the attention more upon this or more upon that one of these complex elements that enter into every moral act of man, and so the recognition of the right or wrong, the corresponding sense of obligation to choose or refuse, and the consequent approval or disapproval, may vary. Hence the consciences of men, however true in themselves, differ in men of different moral habits or dispositions in their estimate of particular action. One's own conscience even varies with his moral mood. The same action is judged and felt by him differently at different times. His intelligence varies in quickness and keenness, and his sensibility in tenderness. But above and beyond this, his will as sovereign may turn the view or the sense now more on one element, now more on another. Even one's own conscience is not uniform in its action.

Nevertheless conscience remains to man the highest arbiter and ruler in all his moral life. The authority of the divine ruler and judge speaks only through that. If the human conscience is not infallible, it is yet the supreme arbiter within the man himself in all morality. Man knows no higher in any department of his nature. The will itself in all its sovereignty must yield to the arbitrament of conscience ; for the creator has not with freedom granted exemption from responsibility. As the mind by the necessities of its nature, must be conscious of its own action, so the will must to some degree at least, pass its own determinations in review before the censorship of the conscience. It may to some extent hinder, or defer, or even mar the action of conscience ; but it cannot wholly silence nor so corrupt as to destroy it.

Hence arises the duty and the importance not only of training and cultivating the conscience, but also of securing it from being stifled or warped by a perverse will or by any particular occasion for its action.

CHAPTER VI.

HOPE, FAITH, AND LOVE.

Virtues.
§ 220. HOPE, FAITH, and LOVE are not only three comprehensive graces; they are also comprehensive virtues.

They sometimes appear with the sensibility, the feeling, predominant and so characterizing them, and then consequently are proper graces. They sometimes, however, appear with the moral element, the free will, predominant in them and so characterizing them as virtues.

As graces they come but indirectly, while as virtues they come directly, under this law; but under the law of duty in both cases they are properly subjects of immediate command. The practical reason, the conscience, recognizes them as right, as obligatory, as praiseworthy, and accordingly by its voice of authority as the organ of the divine will and word, commands them. As graces, they appear characteristically as spontaneous; as virtues they appear as voluntary and free. As thus enjoined duties, in these exercises the will puts itself forth and embodies itself in the feeling as its needful body and form of expression. It leads the feeling to its object; keeps the feeling on its object and animates it to its proper degree of life and tenderness, and moreover protects it from being smothered or overpowered by any adverse feeling.

§ 121. IN HOPE, the free will leads the feeling of desire fed with expectation to its proper object. This object, as legitimate to the human soul, is good, and in order to hope as an enjoined virtue the good hoped for must be the highest good which is possible in the case.

<small>Hope explained.</small>

Hope, as a virtue, may be defined to be the choice of good as the object of desire and expectation.

Hope as an enjoined duty and virtue comprises several leading distinguishable elements and modifications which we proceed to enumerate.

1. Hope as a duty implies something positive to be done. It is not a wholly passive exercise, a mere grace. The will is summoned to go out and find the proper object of hope and put the feeling in exercise. Such object in some form is ever attainable. As surely as the activity of the soul was ordained and fashioned and conditioned in infinite wisdom and goodness for good as its end, so surely is it that the duty of hope is a practicable one under the rule of God. The good in the nature of things connected with right action, is in the duty of hope to be sought and proposed as object to the sensibility.

<small>1. Implies something to be done.</small>

2. In the duty of hope, the desire and expectation are to be set on this good by the sovereign direction of the will.

<small>2. Directed by the will.</small>

3. The duty of hope is both generic and specific. The whole activity of the soul is to be subject to hope in such sense that each governing purpose or choice shall be inspired by it; the whole man is to move on in hope. And subordinate volitions are to stand in like relation to the duty of hope, receiving each its special inspiration from it. No duty can be rightly and perfectly discharged except as thus inspired by hope.

<small>Generic and specific.</small>

4. Hope has its limitations both as to the kind of its objects and the degree of its allowance. The one legitimate object of hope in its generic and supreme exercise, is the good for which the soul was designed and fashioned. The will is enjoined in this duty to seek out and choose this good as highest object of desire and expectation. The duty prohibits any other good to be thus taken as the object of the soul's governing hope. Among the objects of specific hope there is wide room for selection. Some objects are absolutely prohibited; other objects are prohibited only because in the circumstances less worthy than others which are presented or may be found.

The highest legitimate good brings no limitation to hope in degree but such as is imposed by the capacity of the soul itself or by the due demands of other capacities in its culture and regulation. Allowable specific objects of hope are limited in their demands to their due measure of desire and expectation. These limitations vary indefinitely with condition and circumstance.

5. Finally the free will is enjoined in the duty of hope not only to find its proper object and regulate it to its proper degree, but also to guard and protect it from being overborne, and also to sustain and nourish it, that as participating in an active living nature it may ever grow and strengthen.

To be sustained by the will.

§ 222. IN FAITH, the free will leads the natural feeling of dependence to its proper object.

Faith defined.

Faith, as a duty, may accordingly be defined to be the choice of the proper object of dependence. It involves the actual exercise of this feeling in reliance and trust.

The objects of faith are all those objects on which man may in any way depend.

Its object.

Its highest form is in relation to God, as the creator and disposer of man. The office of faith in this its highest form, is to recognize God as the one comprehensive, legitimate, absolute ground of dependence and trust. In this highest form, faith is well characterized as "the subtle chain that binds us to the infinite." In lower and subordinate forms, faith finds its legitimate specific objects in all the beings within its reach which fill the universe of God and in all the events of his providential rule. Especially does it find legitimate objects in the fellow-beings of the same rational nature. Manifold modes and degrees of dependence determine manifold forms and measures of faith. Even the manifold capacities and functions of the soul itself call for manifold kinds and measures of faith as they are interlocked with one another in manifold forms and degrees of reciprocal interdependence.

Faith, as a duty, like hope involves divers elements and modifications. It implies something positive to be done; it involves the fixing of the feeling of dependence necessarily belonging to a finite nature on its proper object or ground, whether this object or ground is the highest and most comprehensive as God himself, or subordinate as his creatures and ordinances; it has its limitations both as to object and degree; and requires protection and nourishment.

Its elements.

§ 223. IN LOVE the free will leads out the natural feeling of sympathy to its proper object.

Love defined.

Love, as a duty, may accordingly be defined to be the choice of the proper object for sympathy. It involves the actual exercise of this sympathy towards its object.

The sphere of love as a duty to man, is commensurate with the range of human sympathy. With whatever being the human soul can be in sympathy and in whatever way such sympathy can be felt and manifested, towards that being and in that way the duty of love extends.

Its sphere.

Its highest forms are in relation to those objects or beings with which the soul is in closest, broadest, deepest relations of sympathy. No being is so near to the soul as its creator and disposer. No being can engage or reciprocate such deep sympathies.

Love consequently is highest and most imperative towards him. It is supreme and comprehensive of all exercises of love towards inferior beings.

As there can be nothing more worthy to engage our sympathy, nothing in a particular being that is more worthy to enlist our highest and warmest sympathy, than the comprehensive good for which he exists, so love in its highest and most commanding form involves sympathy with this end for which the object has his being. If we reverently characterize the end of God's being as his infinite blessedness or the perfect glory of his character, then our love to him must necessarily express sympathy with this end as its highest possible form. Love to God thus in its highest form is will to please him or will to glorify him. As the end of man's being is his blessedness or true excellence of character, love to man in its highest, most generic form, is will to promote this well-being in him.

The specific and subordinate forms of love respect the manifold specific attributes and relations and conditions of other beings so far as they can enlist our sympathy.

Love, as a duty, like hope and faith, involves divers elements and modifications. It

Divers elements. implies a positive act of will, something to be done; it involves the fixing of the natural sympathy of the soul on its appropriate object in kind and allowing to its natural expression its proper degree; it requires protection and nourishment as being subject to culture and growth.

INDEX.

Action, attributes of, 183
Activity, especial attribute of mind, 5-8; peculiar to mind, 6; three forms of mental activity, 7
Æsthetic Sense, 75
Affections, 85-87; defined, 85; classed, 86; modifications, 86
Apparitions, in part accounted for, 65
Appetites, 90
Artistic Imagination, 161
Association of ideas, 147; laws, 147; general principle, 150; special laws, 151
Attention, defined, 211
Attributes, intrinsic or extrinsic, 182
Aversions, 89
Beautiful, emotion of the, 78; its effect, 79; category of, 197
Being, ambiguities of term, 200
Catalepsy, 123
Category, defined, 181; systems of categories of thought, 184-198
Cause, origin of idea, 204
Comic, 80
Choice, 214-223; an act, 215; appropriative, 215; free, 216; personal, 217; sovereign, 217; originative, 220; moral, 221; responsible, 222
Comprehensive knowledge, 208
Concept, how formed, 180
Conscience, explained, 234; synonyms, 234; threefold element, 234; subject to the will, 238
Condition, attributes of, 183
Conscious subject, 35
Continuousness of mind, 22-29
Copula, element of thought, 178
Curiosity, defined, 210
Demonstration, 209
Desires, 88-93; defined, 88; classified, 88
Determination, a logical process, 181
Discursive intelligence, 179
Dreaming, 171
Emotions, 74-84; defined, 74; their classes as awakened by the true, the beautiful, and the good, 75; modifications, 76
Emulation, 92
Entity, 200
Exalted sensibility, 114

Existences, 199-207; origin of notion, 199; classes, 203
Experience, source of knowledge, 1
Extrinsic Attributes, 182-3
Faith as a virtue, 241; defined, 244; its object, 244; elements, 245
Feelings, classified, 54
Form, 53
Generalization, 181
Good, category of, 197
Habit, its nature, 25; condition of growth, 28
Hearing, source of, 76
Hope as a virtue, 241; defined, 242; implies something to be done, 242; directed by the will, 242; genuine and specific, 243; to be sustained by the will, 244
Hope and Fear, 93
Idea, nature of, 35-45; sole object for the mind, 36; of mind and for mind, 37; defined, 39; threefold, 39; as true, 41; as beautiful, 42; as good, 42; respective objects for threefold functions of the mind, 45
Identity, personal, 24
Ideals, 103: primitive and secondary, 104
Ideality, category of, 187
Imagination, faculty of form, 53; defined, 100; artistic, philosophical, and practical, 159-166
Intellectual sense, 75
Intellectual apprehension and representation, 208
Intelligence, function of mind, 7; defined, 164; modifications, 164
Intrinsic attributes, 182; qualities and actions, 183; essential or accidental, 183
Intuition, 167; defined, 172; synonym, 172; sphere, 173; a presentative knowledge, 173; an immediate knowledge, 173
Knowledge, its source, 1; through attributes, 3
Love, as a virtue, 241; defined, 244; sphere, 245; elements, 246
Ludicrous, emotion of the, 79
Memory, 26; attaches to feelings, thoughts, and purposes, 26-28;

INDEX.

	PAGE.
defined, 132; its perpetuity, 134; proved from presumption, 134; analogy: 134; ordinary experience, 135; facts of extraordinary experience, 137; conditions of a good memory, 141; rules,	144
Mental reproduction,	146-158
Mind, synonym of soul, etc., 1; general attributes, 5; essentially active, 5; defined, 6; single and simple, 9-12; distinct from its object, 10; finite and dependent, 13-16; limited in range, 13; in intensity, 13; in object, 14; dependent on objects, 14; on channels, 15; limited in control of object, 16; its passivity, 17-21; from other objects and from its own states, 17; active and passive—a faculty and a capacity—at same time, 19; activity attended with pleasure, 19; continuous, 22-29; self-conscious, 30-34; its relationship, 35-45; sole object for idea, 36; symbols, 46-50; reality proved,	203
Modality, category of,	192
Moral sense, 75; how awakened,	81
Motion, origin of idea,	203
Motive, 224-228; defined, 224; a good, 225; to the whole mind, 226; in the mind, 227; classes,	227
Nerves, sensitive and motor,	64
Observation,	211
Organic sense,	68
Passions,	99
Passivity of mind,	17-21
Perception, 176; defined, 168; the relations to sensation, 168; sphere, 171; a presentative knowledge,	171
Phantoms,	111
Philosophical imagination,	161
Pleasure and pain, 56-61; degrees, 57-58; enter into all mental states, 58; modifications, 59; bodily,	66
Practical sentiments, 97; imagination,	162
Predicate, element of thought,	178
Presentative knowledge,	165
Properties, intrinsic attributes, 183; category of properties and relations,	193
Psychology, defined, 1; its province, 2; method of study,	3-4
Qualities,	183
Quantity, category of,	190
Rational desires, of freedom, power, knowledge,	91, 92
Reasoning, 181; mediate and immediate,	182
Recollection, 156; rules,	159
Redintegration, Hamilton's law of	148
Reflection,	211

	PAGE.
Relation, attributes of,	183
Representative knowledge,	165
Reproduction, 146-158; defined, 146; spontaneous in revery or voluntary in recollection,	146
Resentments,	87
Revery,	147
Self, synonym of mind, etc.,	1
Self consciousness, 30-34; its immediate object, 31; knowing and feeling, 31; degrees,	31
Self love,	89
Sensations, 62-73; defined, 62; medium, 63; classes,	66
Sense ideals, 107-127; defined, 107; modifications,	180
Sensibility, function of mind, 7; notion and modifications, 51-55; capacity of form,	54
Sentiments, 94-97; defined, 94; classes,	95
Sight, sense of,	72
Singleness and simplicity of mind	9-12
Sleep,	115
Smell, sense of,	71
Social desires,	93
Somnambulism,	124
Soul, synonym of mind, etc.,	1
Space, origin of idea,	205
Special senses,	69
Spirit, synonym of mind, etc.,	1
Spiritual ideas, 128-131; defined, 128; source, 128; bodied in ideas,	130
Subject, element of thought,	178
Sublime, emotion, of the,	78
Substance and cause, category of, 195; origin of idea of an existence,	204
Suspended sensibility,	116
Symbols of mind,	46-50
Taste, sense of,	70
Thought, defined, 175; synonyms, 175; follows perception and intuition, 177; its elements, 178; three genuine forms;—the judgment, the concept, and the reasoning,	180
Time, origin of idea,	205
Touch, sense of,	67
True, beautiful, and good, categories of,	197
Universe, origin of idea of,	204
Vital sense,	68
Volition, see under *choice,* predominant and subordinate,	23
Will, function of mind, 7-8; its nature and modifications, 212; synonyms, 212; in choice, 214; growth and subordinations, 229-233; dependent, 230; self-ruling,	231
Wit,	77

www.ingramcontent.com/pod-product-compliance
Lightning Source LLC
Chambersburg PA
CBHW021344230426

43666CB00006B/396